Studies in Systems, Decision and Control

Volume 548

The series "Studies in Systems, Decision and Control" (SSDC) covers both new developments and advances, as well as the state of the art, in the various areas of broadly perceived systems, decision making and control–quickly, up to date and with high quality. The intent is to cover the theory, applications and perspectives on the state of the art and future developments relevant to systems, decision making, control, complex processes and related areas, as embedded in the fields of engineering, computer science, physics, economics, social and life sciences, as well as the paradigms and methodologies behind them. The series contains monographs, textbooks, lecture notes and edited volumes in systems, decision making and control spanning the areas of Cyber-Physical Systems, Autonomous Systems, Sensor Networks, Control Systems, Energy Systems, Automotive Systems, Biological Systems, Vehicular Networking and Connected Vehicles, Aerospace Systems, Automation, Manufacturing, Smart Grids, Nonlinear Systems, Power Systems, Robotics, Social Systems, Economic Systems and other. Of particular value to both the contributors and the readership are the short publication timeframe and the world-wide distribution and exposure which enable both a wide and rapid dissemination of research output.

Indexed by SCOPUS, DBLP, WTI Frankfurt eG, zbMATH, SCImago.

All books published in the series are submitted for consideration in Web of Science.

Fadele Ayotunde Alaba · Alvaro Rocha

Security Framework and Defense Mechanisms for IoT Reactive Jamming Attacks

Springer

Fadele Ayotunde Alaba
Federal College of Education
Zaria, Nigeria

Alvaro Rocha
ISEG, University of Lisbon
Lisboa, Portugal

ISSN 2198-4182 ISSN 2198-4190 (electronic)
Studies in Systems, Decision and Control
ISBN 978-3-031-65928-7 ISBN 978-3-031-65929-4 (eBook)
https://doi.org/10.1007/978-3-031-65929-4

This Springer imprint is published by the registered company Springer Nature Switzerland AG
The registered company address is: Gewerbestrasse 11, 6330 Cham, Switzerland

Preface

The proliferation of Internet of Things (IoT) devices has revolutionized various sectors, including healthcare, transportation and smart cities, by providing unprecedented connectivity and convenience. However, this connectivity comes with significant security challenges, particularly in the form of reactive jamming attacks that can disrupt communications and compromise the integrity of IoT networks. "Security Framework and Defense Mechanisms for IoT Reactive Jamming Attacks" addresses these critical security concerns, offering a comprehensive exploration of the vulnerabilities in IoT networks and the sophisticated tactics employed by adversaries. This book delves into the architecture of IoT systems, the nature of reactive jamming threats and the latest advancements in security frameworks designed to mitigate these risks.

Readers will gain insights into various defense mechanisms, from physical layer security techniques to advanced cryptographic methods and machine learning approaches. The book also provides practical guidelines for implementing these security measures in real-world IoT deployments, ensuring robust protection against reactive jamming attacks. Whether you are a researcher, practitioner, or student in the field of cybersecurity, this book serves as a valuable resource for understanding and defending against one of the most pressing threats in the IoT landscape. Through detailed analysis and expert perspectives, "Security Framework and Defense Mechanisms for IoT Reactive Jamming Attacks" aims to equip you with the knowledge and tools needed to safeguard IoT networks against sophisticated jamming threats.

Zaria, Nigeria
Porto, Portugal
May, 2024

Fadele Ayotunde Alaba
Alvaro Rocha

Acknowledgments

The completion of this book, *Security Framework and Defense Mechanisms for IoT Reactive Jamming Attacks*, would not have been possible without the support and contributions of numerous individuals and organizations.

First and foremost, we extend our deepest gratitude to our families for their unwavering support and understanding throughout this journey. Their encouragement has been invaluable.

We would like to express our sincere appreciation to our colleagues and collaborators for their insights, feedback and collaborative efforts. Their expertise and perspectives have significantly enriched the content of this book.

Special thanks are due to our research institutions and funding bodies, whose support has been instrumental in advancing our research and making this publication possible. Their commitment to fostering innovation in the field of cybersecurity has been a constant source of motivation.

We are also grateful to the technical reviewers and editors whose meticulous work ensured the clarity and accuracy of this book. Their dedication to maintaining high standards has greatly improved the quality of our work.

Finally, we wish to acknowledge the contributions of the broader cybersecurity community. The shared knowledge and ongoing discussions within this community have been a continual source of inspiration and learning.

This book is dedicated to all those who strive to make the world of IoT safer and more secure. Thank you for your commitment to this vital field.

Zaria, Nigeria — Fadele Ayotunde Alaba
Porto, Portugal — Alvaro Rocha

Contents

1 Introduction ... 1
1.1 Background ... 1
1.2 Motivation ... 2
1.3 Objectives ... 3
1.4 Contributions ... 4
1.5 Book Outline ... 5
References ... 6

2 Internet of Things ... 9
2.1 Internet of Things ... 9
2.1.1 Conventional Security Versus Security in IoT ... 10
2.2 Enabling Communication Technologies for IoT ... 13
2.2.1 Wireless Sensor Network (WSN) ... 13
2.2.2 Radio Frequency Identification (RFID) ... 14
2.2.3 Wireless Fidelity (WiFi) ... 14
2.2.4 The 4G, Long-Term Evolution (LTE), LTE-Advanced (LTE-A) and 5G ... 14
2.3 Security Issues of IoT Communication Technologies ... 15
2.3.1 WSN ... 15
2.3.2 RFID ... 17
2.3.3 WiFi ... 18
2.3.4 The 4G LTE, LTE-Advanced (LTE-A) and 5G ... 18
2.4 IoT Communication Channel: Challenges ... 19
2.4.1 MitM Attacks ... 19
2.4.2 Eavesdropping ... 20
2.4.3 Counterfeiting Attacks ... 20
2.5 Security Issues in IoT Network ... 20
2.5.1 Architecture Security ... 21
2.5.2 Authentication ... 24
2.5.3 Heterogeneity ... 28
2.5.4 Quality of Service (QoS) ... 29

2.5.5 Trust Management (TM) 30
2.5.6 Jamming 31
2.6 Security Approaches 35
2.6.1 Lightweight Approaches 35
2.6.2 Model-Based Security Approaches 37
2.7 Implementation Techniques of Existing Approaches for Preventing Jamming Attacks 38
2.7.1 Software Attack Implementation Techniques for Jamming Attacks 38
2.7.2 Hardware Attack Implementation Techniques for Jamming Attacks 39
2.8 Existing Security Countermeasures for Detecting Jamming in IoT Network 40
2.8.1 Existing TJC 40
2.9 Components for Developing IoT Security Framework 41
2.10 Gap in Research 42
2.11 Conclusion 44
References 46

3 Embedded IoT Security Framework 53
3.1 System Model 54
3.2 Basic Factors for Improving IoT Security Framework for Reactive Jamming 55
3.3 Proposed Embedded Security Framework and Architecture 57
3.4 Proposed Authentication and Encryption Scheme for IoT Framework 60
3.5 GCM Notation 61
3.5.1 Encryption 61
3.5.2 Decryption 62
3.5.3 GF (2128) Multiplication 64
3.5.4 The Proposed Protocol 64
3.6 Operational Process for the Proposed Protocol 66
3.6.1 Using AES-GCM 68
3.7 Reactive Jamming Attack Modelling 69
3.8 Activity Model for Reactive Jamming Attacks 70
3.8.1 Reactive Jamming Activity Modelling 71
3.9 Behavioural Model for Reactive Jamming Attack 72
3.9.1 Behavioral Model for Reactive Jamming Attack 73
3.10 Reactive Jamming Model Formulation 74
3.11 Conclusion 75
References 76

4 Defense Mechanism for Reactive Jamming Attack in IoT Networks 79
4.1 The Proposed CDCA for IoT 79
4.1.1 CDCA Network and Attacker Assumptions 80
4.1.2 The Proposed CDCA 81
4.2 Simulation of CDCA 85
4.2.1 Implementation Details 85
4.3 Results and Discussions 85
4.3.1 Performance Results by Varying Interval 86
4.3.2 Performance Result Under Malicious Nodes 88
4.3.3 CDCA Performance Result Under Realistic Conditions 91
4.3.4 Performance of CDCA Under Mobility 93
4.4 Conclusion 95
References 96

5 Statistical Results Validation 97
5.1 Statistical Models 97
5.2 Coefficient of Determination 99
5.3 Validation of Results 100
5.3.1 Analysis of the Results Regarding Energy Consumption 100
5.3.2 Analysis of the Results Regarding Delay 103
5.3.3 Analysis of the Results Regarding Delay Throughput 106
5.4 Conclusion 110
References 110

6 Conclusions and Future Directions 111
6.1 Conclusion 111
6.2 Significance 112
6.3 Future Directions 113

Symbols and Abbreviations

4G	Fourth Generation
5G	Fifth Generation
6LowPANs	IPV6 over Low-power Wireless Personal Networks
ABE	Attribute-Based Encryption
AES	Advanced Encryption Standard
ARM	Architecture Reference Model
ASICs	Application Specific Integrated Circuits
ASM	Activity Sequential Modelling
AWFQ	Adaptive Weighted Fair Queue
BS	Base Station
CDCA	Countermeasure Detection and Consistency Algorithm
CLC	Certificate-Less Cryptography
CoAP	Constrained Application Protocol
CPU	Central Processing Unit
DoS	Denial of Service
DST	Digital Signature Technology
DTLS	Datagram Transport Layer Security
E2E	End-to-End
ECCDH	Elliptical Curve Cryptography Diffie Hellman Algorithm
ECDSA	Elliptic Curve Digital Signature Algorithm
EPC	Electronic Product Code
FWQ	Fuzzy Weighted Queuing
GCM	Galois Counter Mode
GPP	General Purpose Processor
H2H	Human-to-Human
H2T	Human-to-Things
HTTP	Hypertext Transport Protocol
IBC	Identity-Based Cryptography
ICT	Information Communication Technology
IoT	Internet of Things
IPv6	Internet Protocol version 6

IT	Information Technology
LLNs	Low-power Lossy Networks
LTE	Long Term Evolution
M2M	Machine-to-Machine
MAC	Message Access Control
MCU	Microcontroller Unit
MIMO	Multiple-Input Multiple-Output
MitM	Man-in-the-Middle
MU-BF	Multiple Users Beamforming
OFDM	Orthogonal Frequency Division Multiple
OSCAR	Object Security Architecture
PCs	Personal Computers
PGP	Pretty Good Privacy
PHY	Physical Layer
PKE	Public Key Encryption
PKIs	Public Key Infrastructures
PRE	Probabilistic Resource Estimation
QoS	Quality of Service
RBAC	Role-Based Access Control
REAC	Regressive Admission Control
RSU	Road Side Unit
SDIoT	Software Defined Internet of Things
SDN	Software Defined Network
SDSec	Software Defined Security
SDStore	Software Defined Storage
SEA	Symmetric Encryption Algorithm
SEE	Secure Execution Environment
SSO	Single Sign-On
SU-BF	Single User Beamforming
T2T	Things-to-Things
TCP	Transmission Control Protocol
TM	Trust Management
TPM	Trusted Platform Modules
UID	User Identification
Wi-Fi	Wireless Fidelity
WSN	Wireless Sensor Network

List of Figures

Fig. 2.1 Landscape of IoT ... 11
Fig. 2.2 Communication flow in IoT environment ... 11
Fig. 2.3 IoT network versus conventional network ... 12
Fig. 2.4 Classification of existing authentication schemes for IoT ... 27
Fig. 2.5 Functional model for constant jammer ... 33
Fig. 2.6 Functional model for deceptive jammer ... 34
Fig. 2.7 Functional model for random jammer ... 34
Fig. 2.8 Functional model for reactive jammer ... 35
Fig. 3.1 Reactive jamming model ... 54
Fig. 3.2 IoT attack communication scenario using use case ... 55
Fig. 3.3 Reactive jamming attacks security modelling ... 56
Fig. 3.4 Basic requirements for security framework for IoT ... 58
Fig. 3.5 Proposed security framework for embedded IoT ... 59
Fig. 3.6 Data encryption operational process ... 62
Fig. 3.7 Data decryption operational process ... 63
Fig. 3.8 Capability structure for IoT devices ... 65
Fig. 3.9 Proposed protocol ... 67
Fig. 3.10 AES-GCM encryption and authentication process for IoT devices ... 68
Fig. 3.11 AES-GCM decryption and authentication process for IoT devices ... 69
Fig. 3.12 System model for sender node, receiver node, and reactive jammer ... 70
Fig. 3.13 Activity model for reactive jammer ... 71
Fig. 3.14 Sequential model for reactive jamming attack ... 73
Fig. 4.1 CDCA flow chart ... 83
Fig. 4.2 IoT network states with CDCA and reactive Jamming ... 83
Fig. 4.3 CDCA working process ... 84
Fig. 4.4 Ave. energy consumption for CDCA, TJC, and reactive jamming traditional technique under varying traffic interval ... 86

Fig. 4.5 Delay for CDCA, TJC, and reactive jamming traditional technique under varying traffic interval 87
Fig. 4.6 Throughput for CDCA, TJC, and reactive jamming traditional technique under varying traffic interval 88
Fig. 4.7 Ave. energy consumption for CDCA, TJC, and reactive jamming traditional technique under malicious nodes 89
Fig. 4.8 Delay for CDCA, TJC, and reactive jamming traditional technique under malicious nodes 89
Fig. 4.9 Throughput for CDCA, TJC, and reactive jamming traditional technique under malicious nodes 90
Fig. 4.10 Ave. energy consumption for CDCA, TJC and reactive jamming traditional technique under realistic condition with malicious nodes 91
Fig. 4.11 Delay for CDCA, TJC and reactive jamming traditional technique under realistic condition with malicious nodes 92
Fig. 4.12 Throughput for CDCA, TJC and reactive jamming traditional technique under realistic condition with malicious nodes 92
Fig. 4.13 Ave. energy consumption for CDCA, TJC and reactive jamming traditional technique under mobility with malicious nodes 93
Fig. 4.14 Delay for CDCA, TJC and reactive jamming traditional technique under mobility with malicious nodes 94
Fig. 4.15 Throughput for CDCA, TJC, and reactive jamming traditional technique under mobility with malicious nodes 94

List of Tables

Table 2.1 IoT security architectures and application domain 21
Table 2.2 Functional evaluation of existing frameworks 42
Table 2.3 Comparison of existing approaches for jamming based on functional optimization and countermeasures attacks 43
Table 3.1 NGF (2^{128}) multiplication 64
Table 3.2 Notations and descriptions 66
Table 4.1 Flow of CDCA 81
Table 5.1 Summary of the purpose of analysis energy consumption, delay and throughput 98
Table 5.2 Comparison between the CDCA, TJC and traditional techniques in terms of energy consumption under varying traffic interval 100
Table 5.3 Comparison between the CDCA, TJC and traditional techniques in terms energy consumption under malicious nodes 101
Table 5.4 Comparison between the CDCA, TJC and traditional techniques in terms energy consumption under realistic condition with malicious nodes 101
Table 5.5 Comparison between the CDCA, TJC and traditional techniques in terms energy consumption under mobility with malicious nodes 101
Table 5.6 Predefined fit function using multiple regressions regarding energy consumption 102
Table 5.7 Analysis of multiple regression for CDCA and the State-of-the-arts approaches with respect to energy consumption under of malicious nodes 102
Table 5.8 *t*-test significance of difference between the CDCA and the State-of-the-arts approaches on energy consumption 103
Table 5.9 Evaluation comparison: CDCA, TJC and traditional techniques regarding energy consumption 103

Table 5.10 Comparison between the CDCA, TJC and traditional techniques in terms traffic delay under varying traffic interval 104
Table 5.11 Comparison between the CDCA, TJC and traditional techniques in terms traffic delay under malicious nodes 104
Table 5.12 Comparison between the CDCA, TJC and traditional techniques in terms traffic delay under realistic condition with malicious nodes 104
Table 5.13 Comparison between the CDCA, TJC and traditional techniques in terms traffic delay under mobility with malicious nodes 104
Table 5.14 Predefined fit function using multiple regressions regarding delay 105
Table 5.15 Analysis of multiple regression for CDCA, TJC and traditional techniques in terms under of malicious nodes 105
Table 5.16 *t*-test significance of difference between CDCA, TJC, and traditional techniques on delay 106
Table 5.17 Evaluation comparison: CDCA, TJC, and traditional techniques concerning delay 106
Table 5.18 Comparison between the CDCA, TJC, and traditional techniques in terms of traffic delay under varying traffic interval 107
Table 5.19 Comparison between the CDCA, TJC and traditional techniques in terms traffic delay with normal nodes 107
Table 5.20 Comparison between the CDCA, TJC and traditional techniques in terms traffic delay under realistic condition with normal nodes 107
Table 5.21 Comparison between the CDCA, TJC and traditional techniques in terms traffic delay under mobility with malicious nodes 107
Table 5.22 Predefined fit function using multiple regression regarding throughput 108
Table 5.23 Analysis of multiple regression for throughput of CDCA and the contemporary schemes 108
Table 5.24 *t*-test significance of difference between CDCA and the contemporary Schemes with respect to the throughput 109
Table 5.25 Evaluation, comparison: CDCA, TJC and traditional techniques in terms of throughput 109

Chapter 1
Introduction

This chapter presents the background, motivation, objective, contributions and organization of this book.

1.1 Background

The Internet-enabled devices/computers are now embedded in everything from cars to pacemakers and from electronic door locks to home heating controllers, and equipped with the ability to communicate with one another via the wireless medium, forming a network of Internet of Things (IoT). Wireless Sensor Network (WSN) is the main technology used in developing these devices; an example is the Internet of Things (IoT) device. The IoT has recently emerged as an important research topic. IoT, a pervasive fabric that interconnects many machines and devices, is one of the most critical technologies of the coming decade [1]. IoT is regarded as the future extension of the Internet [2]. The IoT comprises physical objects such as sensors, actuators, mobile phones and Radio Frequency Identification (RFID) tags, which can sense, monitor, communicate, and exchange data with each other to perform different tasks around a specified location [3]. IoT provides an interconnection of various types of devices over possibly vast heterogeneous networks so that the devices can communicate directly with each other without human intervention. Recently, IoT has gain popularity because of "increasing applications" that connected to the Internet and researchers [4–6] predicted that the number of connected things by year 2020 will be four (4) times more than the human population. The arrival of IoT brings the constant universal connection of people, obj. Its services and main objective are to always provide a network infrastructure with interoperable communication protocols on any network [7]. The growth and development of smart technology allow numerous devices and objects to be a part of IoT through different smart sensors.

F. A. Alaba and A. Rocha, *Security Framework and Defense Mechanisms for IoT Reactive Jamming Attacks*, Studies in Systems, Decision and Control 548,
https://doi.org/10.1007/978-3-031-65929-4_1

As large-scale deployment of IoT increases rapidly, it gives rise to major security concerns.

Security is one of the critical attributes of any communication network. Various attacks have been reported over the last many years. Most of them, however, target wired networks. Wireless networks have only recently been gaining wide. With the advances in technology, wireless networks are becoming more affordable and easier to build [8].

Security is one of the critical attributes of any communication network. Various attacks have been reported over the last many years. A major security challenge in IoT is the absence of a unified security architecture and framework to support authentication, heterogeneity, trust management, jamming, etc. [8, 9]. A secure IoT architecture is a significant challenge that needs proper attention for IoT to be fully adopted [10, 11]. The absence of a universally accepted IoT architecture makes IoT devices vulnerable to attacks and threats. Since IoT architecture is expected to manage a projected number of billions of interconnected things, this will create an amazing number of paths accessible by malicious attackers, thus making it more susceptible to jamming attacks.

Also, IoT's heterogeneous nature, which involves billions of devices (having different protocols) to be connected, makes the network vulnerable to attacks because each connected device has additional security measures [7]. Knowing the authentic devices in this vast network will be challenging due to some IoT devices, such as RFID, that do not have access control functions [12]. During communication, smart attackers can easily hijack channel and eavesdrops the sending packets. One of the main constraints of wireless networks is vulnerability against jamming [13]. In [14–17] researches all addresses jamming attacks in WSN as a dangerous attack that block legitimate communication channel between devices by causing intentional interference. With the rapid development of wireless technologies, jamming in IoT networks has become a major research problem due to the ease of blocking communication channels. Before IoT can be accepted and fully deployed securely, there is a need for a robust security framework that can detect jamming attacks and take the necessary recovery actions.

1.2 Motivation

Security is one of the critical features of any communication network. The nature of wireless networks makes them more susceptible to security attacks. The constrains (e.g., limited processing power) inherent in IoT devices limit their ability to defend themselves from attacks. The IoT devices are more vulnerable to jamming attacks, which disrupt their data transmission, causing them to re-transmit repeatedly. This, in turn, quickly drains their battery, hindering operations in an IoT network and negatively affecting performance. Jamming could lead to a denial of service (DoS) attack, the most common attack in WSNs and the Internet [18]. Although researchers have proposed frameworks that support various IoT-constrained devices based on identity certificate management, single sign-on, federated identity, and user-centric framework, each framework has its limitations in terms of functional performance. For example, Huston et al. in [19] and Guneysu and Oder in [20] introduced Public Key Infrastructure and Pretty Good Privacy (PGP), both using identity certificate management. However, the frameworks do not support functions like single sign-on, federated identity, user-centric, and device security. Similarly, Newman and Tsó in [21] introduced Kerberos using single sign-on, but the framework does not support other functions like identity certificate management, federated identity, user-centric, and device security. In addition, Chehab and Abdallah in [22] introduced the liberty alliance. The framework has functional features such as single sign-on, federated identity, and user-centric but does not support identity certificate management or device security functions. Note that there is no framework for device security among all the existing frameworks.

In addition, different researchers have proposed various countermeasures for reactive jamming. A novel anti-jamming strategy was proposed by Namvar et al. [8] to prevent IoT devices from malicious radio jammers. The anti-jamming scheme uses Colonel Blotto's game theory, which lacks good energy consumption. Wood et al. in [23] introduced energy-efficient routing for jamming attacks. This approach manages energy at the node by adopting frequency hopping. However, it causes an increase in traffic delays and consumes more energy. Jamming attack detection and countermeasures using Ant System to prevent and detect jamming attacks in IoT networks was proposed by Heo et al. [24]. The challenge of this approach is an increase in delay. However, based on the challenges and drawbacks of both the security frameworks and jamming attack detection countermeasures, IoT demands novel security methods that will help in threat monitoring and detecting jamming attacks. Securing resource-constrained devices remains one of the most challenging issues that connected devices need to address.

1.3 Objectives

This book embarks the following objectives.

a. To analyze and identify the limitations of existing security frameworks and techniques for detecting jamming attacks.
b. To develop a security framework to detect reactive jamming attacks.
c. To design a behavioural model for reactive jamming attacks using the Unified Modelling Language and formulate a mathematical model for reactive jamming using game theory.
d. To develop a defense mechanism for reactive jamming on IoT using a novel Countermeasure Detection and Consistency Algorithm (CDCA) for the reactive jamming attack.
e. To evaluate the defense mechanism using the state-of-the-art countermeasure approaches.

1.4 Contributions

We studied the state-of-the-art frameworks and countermeasures for jamming attacks in IoT networks to identify the issues and drawbacks in the existing frameworks and countermeasures against jamming attacks for IoT networks. A thematic taxonomy that categorizes the existing frameworks and jamming countermeasures is proposed, and a qualitative comparison is carried out between them. All contemporary frameworks address different security issues, but none address security for resource-constrained devices. A detailed understanding of the existing solutions leads to the development of an authentication and capability-based device security framework for IoT networks using "Advanced Encryption Standard (AES)-Galois Counter Mode (GCM)" (a security protocol for constrained IoT devices). The AES-GCM is an encryption process that is based on authentication. The AES-GCM is suitable for IoT, because of its high-speed encryption approach and low computational latency, which is useful for devices with high data rates like IoT devices [25]. Also, when there is no data to encrypt, it acts as a stand-alone Message Access Control (MAC) just for message authentication [26].

In addition, the proposed protocol is evaluated by introducing reactive jamming attacks on the communication channel. We investigated the reactive jamming attack and its impacts on IoT networks. It was observed that among all the existing jamming attacks, reactive jamming is the most disastrous type of jamming attack due to its intelligent behavior. Based on this, we formulated a reactive jamming model in this research and developed a novel Countermeasure Detection and Consistency Algorithm (CDCA) for detecting reactive jamming attacks in IoT networks. The proposed CDCA uses a threshold-based mechanism to detect and cure the attack. In this case, every node maintains some send threshold value and periodically compares the current transmission with the threshold. The proposed CDCA technique uses a change in the sending threshold. The proposed technique also boosts IoT network performance (packet signal strength) and improves the location consistency checks of connected devices, thereby protecting the network against reactive jamming.

The algorithm uses channel signal strength to check for packet consistency when there is a jamming attack along the communication channel by checking if the data transmission value contradicts the sending data threshold. The algorithm sending the threshold value is periodically checked and compared with the current value after data transmission, to determine if there is an attack on the network. Also, jamming attack is detected when the signal strength is high. However, if the signal strength is low, the node may or may not be jammed.

When the defense mechanism detects the jamming node, it notifies all other connected nodes about the jammer node, changing all paths coming from the jammed node. As a result, the defense mechanism indirectly removes the jammer node from the network. The jammer node is removed completely through the isolation process. That is, the defense mechanism isolates the jammer node path by changing the direction of data transmission among normal nodes in the network. This process will prevent the jammer node from transmitting and drain the little energy on the node, thus disconnecting itself from the network. The performance of the proposed CDCA is evaluated using Cooja simulator. Using realistic simulation scenarios (e.g., with varying traffic interval, number of different malicious nodes, and random mobility patterns), the simulation results demonstrate the performance supremacy of CDCA compared to contemporary schemes in terms of various performance metrics, such as energy consumption, traffic delay, and network throughput.

The main benefit of CDCA is that its defense mechanism supports increasing the number of jamming nodes in the network and can cure reactive jamming attacks. We adopted game theory as the defense mechanism for the proposed reactive jamming attack. Finally, CDCA is validated using statistical models.

Firstly, this book presented a comprehensive survey of the state-of-the-art IoT security threats and vulnerabilities. The security threats and vulnerabilities are discussed regarding application, architecture, and communication. In addition, it examined the current state-of-the-art IoT-enabling communication technologies. This survey's results offer researchers an understanding of security challenges, especially jamming attacks imposed on IoT networks. The result also provided a detail understanding of reactive attacks and their behaviour (impact) on IoT networks and proposed a possible solution. To achieve a sound security solution for embedded IoT devices involves introducing security mechanisms on these devices right from the development stage.

Second, based on the survey results and to achieve a sound security solution for IoT devices, this book proposed a novel security framework and authentication protocol for securing IoT networks. It is a challenge-response type of security framework that improves the computation overhead of IoT devices. The proposed framework provides device security for embedded IoT devices, improving the existing frameworks. Because among all the existing frameworks, there is no framework for device security.

Lastly, it was observed that reactive jamming attacks are difficult to detect in IoT networks due to their behavior across the network. These attacks negatively affect the performance of devices and hinder IoT operations. To address such an issue, this book presents a novel countermeasure and defense mechanism to fight reactive jamming attacks on IoT networks. The proposed defense mechanisms detect and treat attacks using a threshold value change. Based on realistic simulation scenarios (e.g., with varying traffic intervals, number of malicious nodes, and random mobility patterns), the performance of the proposed mechanisms is evaluated. The simulation results demonstrate the superiority of the proposed technique compared with contemporary schemes in terms of performance metrics such as energy consumption, traffic delay, and network throughput.

1.5 Book Outline

The remainder of the book is organized as follows.

Chapter 2 presents the overview of IoT and current related research in securing the IoT application domain. This chapter examines the problems in securing data in IoT networks and the existing solutions to solving jamming attacks. It presents existing solutions that tackle different security issues in IoT networks. Moreover, in the context of the gap, the chapter discussed the current security issues related to existing security frameworks and jamming attacks in IoT networks. The drawbacks and challenges of each work are also stated. Finally, the research's motivation stems from the existing approaches' weaknesses and challenges.

Chapter 3 presents the proposed security framework for IoT networks. The framework was developed using a cost-effective design approach. Two components are used in developing the protocol. The components are Capability Design (mainly a ticket, token, or key that provides authorization to access a device) and Advanced Encryption Standard (AES)-Galois Counter Mode (GCM). This chapter also presents reactive jamming modeling and attack formation in IoT networks. It presents the activity and behavioral (sequential) models that display how messages are exchanged using the Unified Modelling Language (UML) approach. In addition, it also presents reactive jamming attack formulation using a game theoretical approach. Every game has two or more players, the decision-makers, with a set of strategies (actions) available to them. We formulate a reactive jamming attack, a game between two players: "a non-cooperative" and "zero-sum game" to analyze the interaction between the attacker and monitoring nodes in the network.

Chapter 4 presents an efficient defense mechanism for IoT called the Countermeasure Detection and Consistency Algorithm (CDCA) for reactive jamming attacks in IoT networks. Reactive jamming simulations were carried out to compare the CDCA performance to contemporary schemes based on performance metrics.

Chapter 5 presents the results of validating CDCA, TJC, and the traditional techniques using statistical models. Also, it is described by the relationship between the variables to predict the dependent variable from independent variables.

Lastly, Chapter 6 concludes the book by highlighting the significant contributions of this research and future research directions.

References

1. Wei, Z., Zhu, M., Zhang, N., Wang, L., Zou, Y., Meng, Z., et al.: UAV-assisted data collection for internet of things: a survey. IEEE Internet Things J. **9**(17), 15460–15483 (2022)
2. Xu, H., Wu, J., Pan, Q., Guan, X., Guizani, M.: A survey on digital twin for industrial internet of things: applications, technologies and tools. IEEE Commun. Surv. Tutor. (2023). https://doi.org/10.1109/COMST.2023.3297395
3. Dankan Gowda, V., Jacob, S.S., Hussain, N., Chennappan, R., Sakhare, D.T.: Convergence of communication technologies with internet of things. In: International Conference on Intelligent Cyber Physical Systems and Internet of Things, pp. 659–670. Springer International Publishing, Cham (2022)
4. Lu, C.: Overview of Security and Privacy Issues in the Internet of Things, pp. 1–11. (2014)
5. Nolin, J., Olson, N.: The internet of things and convenience. Internet Res. **26**(2), 360–376 (2016). https://doi.org/10.1108/IntR-03-2014-0082

6. Fadele, A.A., Othman, M., Hashem, I.A.T., Alotaibi, F.: Internet of things security: a survey. J. Netw. Comput. Appl. **88**(April), 10–28 (2017). https://doi.org/10.1016/j.jnca.2017.04.002
7. Islam, M.M., Nooruddin, S., Karray, F., Muhammad, G.: Internet of things: device capabilities, architectures, protocols, and smart applications in healthcare domain. IEEE Internet Things J. **10**(4), 3611–3641 (2022)
8. Namvar, N., Saad, W., Bahadori, N., Kelley, B.: Jamming in the internet of things: A game-theoretic perspective. In: 2016 IEEE Global Communications Conference GLOBECOM 2016—Proceedings (2016). https://doi.org/10.1109/GLOCOM.2016.7841922
9. Jing, Q., Vasilakos, A.V., Wan, J., Lu, J., Qiu, D.: Security of the internet of things: perspectives and challenges. Wirel. Netw. **20**(8), 2481–2501 (2014). https://doi.org/10.1007/s11276-014-0761-7
10. Mirani, A.A., Velasco-Hernandez, G., Awasthi, A., Walsh, J.: Key challenges and emerging technologies in industrial IoT architectures: a review. Sensors **22**(15), 5836 (2022)
11. Ang, K.L.M., Seng, J.K.P., Ngharamike, E.: Towards crowdsourcing internet of things (crowd-iot): architectures, security and applications. Future Internet **14**(2), 49 (2022)
12. Sicari, S., Rizzardi, A., Grieco, L.A., Coen-Porisini, A.: Security, privacy and trust in internet of things: the road ahead. Comput. Netw. **76**, 146–164 (2015). https://doi.org/10.1016/j.comnet.2014.11.008
13. Pirayesh, H., Zeng, H.: Jamming attacks and anti-jamming strategies in wireless networks: a comprehensive survey. IEEE Commun. Surv. Tutor. **24**(2), 767–809 (2022)
14. Kasinathan, P., Pastrone, C., Spirito, M.A., Vinkovits, M.: Denial-of-Service detection in 6LoWPAN based internet of things. In: International Conference on Wireless and Mobile Computing, Networking and Communications, no. October, pp. 600–607 (2013). https://doi.org/10.1109/WiMOB.2013.6673419
15. Goyal, D.: Routing protocols in wireless sensor networks: a survey, pp. 478–484 (2012). https://doi.org/10.1109/ACCT.2012.98
16. Babar, S.D., Prasad, N.R., Prasad, R.: Jamming attack: behavioral modelling and analysis. In: 2013 3rd International Conference on Wireless Communications, Vehicular Technology, Information Theory and Aerospace and Electronic Systems. VITAE 2013—Co-located with Global Wireless Summit 2013, pp. 0–4 (2013). https://doi.org/10.1109/VITAE.2013.6617054
17. Mpitziopoulos, A., Gavalas, D., Konstantopoulos, C., Pantziou, G.: A survey on jamming attacks and countermeasures in WSNs. IEEE Commun. Surv. Tutor. **11**(4), 42–56 (2014). https://doi.org/10.1109/SURV.2009.090404
18. Hatzivasilis, G., Papaefstathiou, I., Manifavas, C.: SCOTRES: secure routing for IoT and CPS. IEEE Internet Things J. **73100**(c), 1–1 (2017). https://doi.org/10.1109/JIOT.2017.2752801
19. Huston, G., Michaelson, G., Kent, S.: Resource certification—a public key infrastructure for IP addresses and AS's. In: 2014 IEEE Globecom Workshops Gc Work (2014). https://doi.org/10.1109/GLOCOMW.2009.5360715
20. Guneysu, T., Oder, T.: Towards lightweight identity-based encryption for the post-quantum-secure internet of things. In: 2017 18th International Symposium on Quality Electronic Design, pp. 319–324 (2017). https://doi.org/10.1109/ISQED.2017.7918335
21. Neuman, B.C., Ts'o, T.: Kerberos: an authentication service for computer networks. IEEE Commun. Mag. **32**(September), 33–38 (2014). https://doi.org/10.1109/35.312841
22. Chehab, M.I., Abdallah, A.E.: Architectures for identity management. In: 2009 International Conference for Internet Technology and Secured Transactions {ICITST}, pp. 1–8 (2016). 978-1-4244-5647-5. https://doi.org/10.1109/ISCC.2018.8538636
23. Wood, A.D., Stankovic, J.A., Son, S.H.: JAM: a jammed-area mapping service for sensor networks. IEEE Sens. J. **3**(1), 24–31 (2014)
24. Heo, J., Kim, J., Bahk, S.: Dodge-Jam: anti-jamming technique for low-power and lossy wireless networks. IEEE Access **2**(3), 345–359 (2017)

25. Babar, S.D.: Security framework and jamming detection for internet of things: a dissertation submitted to the Department of Electronic System of Aalborg University in partial fulfillment of the requirements for the degree of doctor of philosophy. [Online]. http://vbn.aau.dk/files/208811916/Thesis_SECURITY_FRAMEWORK_AND_JAMMING_DETECTION_FOR_INTERNET_OF_THINGS.pdf (2015)
26. Hinsenkamp, L., Zhou, G., Michalik, H.: Efficient and high-throughput implementations of AES-GCM on FPGAs. IEEE Des. Test Comput. **12**(5) (2013)

Chapter 2
Internet of Things

This chapter overviews IoT and related research in securing the IoT application domain. The goal is to identify any gap in securing resource-constrained IoT devices. The "gap" is based on the current IoT security approaches for solving the challenges imposed on IoT devices and the technologies that help exchange data between connected devices. Another objective is to find the most recent and relevant solutions to solving the problems of securing highly constrained devices to the Internet. Different solutions are examined and discussed based on the resource requirements and ability to solve the problem fully or partially. Based on this review, comparisons of each solution are made, and research gaps are identified.

2.1 Internet of Things

IoT is the presence of a "variety of objects" around us (e.g., RFID tags, sensors, actuators, mobile phones, etc.), which, through unique addressing schemes, can interact with each other and cooperate with their neighbors to reach a common goal [1]. The potential applications span several fields, such as healthcare, industrial automation, smart homes, smart grid, and smart city [2]. IoT involves the interconnection of vast heterogeneous devices in different communication patterns, for example, human-to-human (H2H), human-to-thing (H2T), thing-to-thing (T2T), or thing-to-things (T2Ts) [3]. One of the characteristics of IoT is the heterogeneity of devices, platforms, wireless technologies, communication protocols, and applications, which form the building blocks of IoT. The diversity of IoT applications and services poses an interesting and difficult challenge that must be addressed to support seamless operation and make possible wide and efficient deployment of IoT shortly. The IoT provides an "integration of various sensors and objects" that can communicate directly with one another without human intervention. The "things" in the IoT include physical devices, such as sensor devices, which monitor and gather all types of data

F. A. Alaba and A. Rocha, *Security Framework and Defense Mechanisms for IoT Reactive Jamming Attacks*, Studies in Systems, Decision and Control 548,
https://doi.org/10.1007/978-3-031-65929-4_2

on machines and human social life [4]. The arrival of IoT has led to the constant universal connection of people, objects, and services. The main objective of IoT is to provide a network infrastructure with interoperable communication protocols and software to allow the connection and integration of physical/virtual sensors, personal computers (PCs), smart devices, automobiles, and household items, such as fridges, dishwashers, microwave oven, food and medicines, anytime and on any network [5]. The development of smartphone technology allows countless objects to be a part of the IoT through different smartphone sensors.

Recently, IoT has become attractive because of the expansion of appliances connected to the Internet. The IoT comprises billions of digital devices/objects, people and services that connect effortlessly to exchange information. The IoT is building a realm for consistently integrating physical items to form an information network with a specific end goal of providing advanced and smart services to the users. The integration involves the combination of embedded systems, sensors, and actuators connected to the universal network. This makes our environment digital (i.e., sensitive, adaptive, and responsive). The connected "things" monitor and collect all kinds of environmental data. It enables the collection of real-time data about the property, individuals, plants, and animals [5]. The IoT is a new idea that involves objects of our daily life, such as automobiles and shopping carts, which will have the capacity to share information and associate with each other and the environment. Along these lines, IoT will enhance a tremendous scope of new industrial opportunities for software and hardware designers [6]. Figure 2.1 illustrates the landscape of IoT with different connected devices, as explained in the definition.

Due to the heterogeneous connections of devices in IoT, as shown in Fig. 2.1 above, security is a major challenge that needs to be addressed to gain users' trust and confidence. The concept of IoT has enticed both academics and industries [7]. IoT is developing at an outstanding speed. The rapid development of IoT has been enhanced through cloud computing due to its characteristics, like provisioning computing resources on demand. The task of computing nodes in IoT are shown in Fig. 2.2. Although IoT is very interesting, it cannot grow if security issues are not resolved [3]. For things to communicate, they must identify each other through enabled communication technologies. At present, the enabling technologies include Wireless Sensor Networks (WSN), Radio Frequency Identification (RFID), Wireless Fidelity (Wi-Fi), ZigBee, 4G LTE, LTE-Advanced (LTE-A), and 5G.

2.1.1 Conventional Security Versus Security in IoT

Several key differences exist between the IoT and conventional wireless networks regarding security and privacy. For example, the deployment of the IoT is unique compared to that of the conventional Internet. IoT devices are set up on low-power and lossy networks (LLNs), whereas conventional networks have fixed or static topologies that rely on the application. LLNs are strained by dynamism, memory, and processing power [8]. These factors are not considered for the standard Internet.

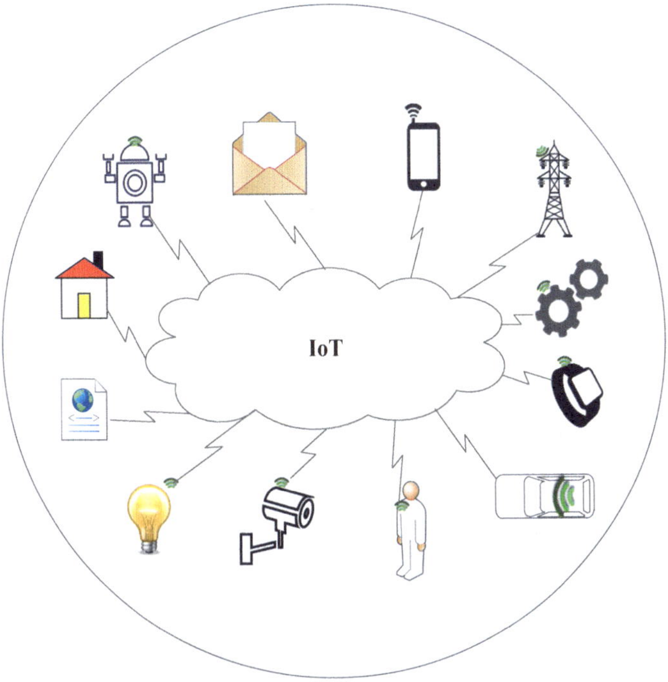

Fig. 2.1 Landscape of IoT

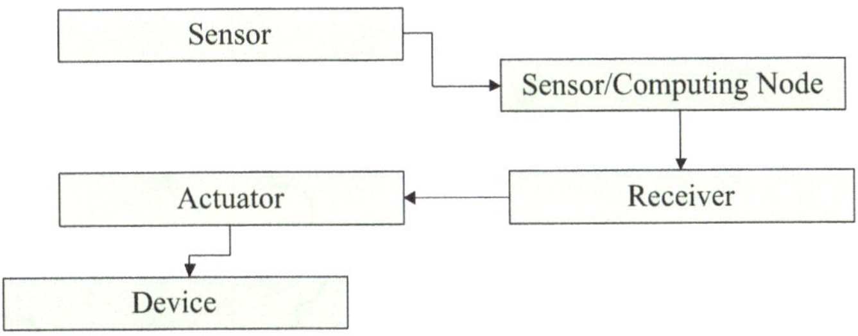

Fig. 2.2 Communication flow in IoT environment

LLNs experience great data losses due to node impersonation [9]. For instance, in the process of data transmission, if an attacker can connect to the network using any identity, the attacker is assumed to be an authentic node. In the case of smart meter applications, the readings can be manipulated by an attacker to send erroneous control messages [8].

The security features and requirements of both the IoT and conventional network devices are also different [4, 10]. In the IoT perception layer, sensor nodes have low computational power and storage capacity, making securing IoT devices using frequency hopping communication application and public key encryption impossible.

Lightweight encryption technology, which includes a lightweight cryptographic algorithm, is used for IoT devices. Among issues at the network layer are man-in-the-middle and counterfeit attacks. Both attacks can capture information and send fake news to communicating nodes in the network [11]. Identity authentication and data confidentiality mechanism are used to prevent unauthorized nodes. At the application layer, data sharing is the main feature. Data sharing creates security challenges in data privacy, access control, and disclosure of information [12]. The security requirements for the application layer include authentication, key agreement, and user privacy protection across heterogeneous networks.

Furthermore, the communication protocols in both conventional and IoT networks differ. Each layer in the network has its communication protocol. For example, IPv6 over low-power wireless personal area networks (6LoWPANs) is used in the IoT perception/physical layer. In contrast, wireless fidelity (Wi-Fi) is used in the physical layer of conventional wireless networks. In the IoT network layer, datagram transport layer security (DTLS) is used as a communication protocol, whereas conventional network uses transmission control protocol (TCP). Constrained Application Protocol (CoAP) is used in the IoT application layer for communication, whereas Hypertext Transport Protocol (HTTP) is used in the application layer of conventional networks [13].

In summary, the conventional security architecture is designed based on users' perspective and is not applicable to communication among machines. The security issues in both networks may be similar, but different approaches and techniques are required to handle each network security issue (Fig. 2.3) [14].

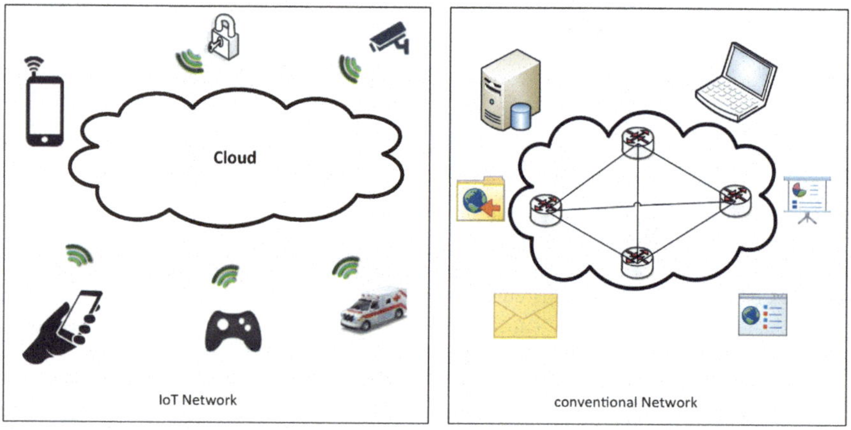

Fig. 2.3 IoT network versus conventional network

2.2 Enabling Communication Technologies for IoT

This section explains the communication technologies that enable IoT devices. It may be short-range communication (for example, Wi-Fi, Bluetooth and ZigBee) or long-range communication (for example, LTE, LTE-A, satellite).

2.2.1 Wireless Sensor Network (WSN)

A WSN is a collection of dedicated sensor devices (an electronic converting device that converts energy from one form to another) with a communication infrastructure that detects the parameters in the environment it monitors. Usually, the monitored parameters include temperature, humidity, pressure, light, vibration and sound quality, power-line voltage, chemical applications, pollutant levels, and fundamental body capacities [15]. A WSN consists of several sensor nodes and a base station (BS) (i.e., sink node). The sensor nodes have limited processing power, constrained battery power, and limited storage capacity, whereas the BS is effective in processing power and storage capacity. Sensor nodes are equipped with many onboard sensors that take readings (e.g., voltage) from their environment. They collaborate and convey the sensed data to the BS in an ad hoc manner. A commonly used commercial WSN solution is based on the IEEE 802.15.4 standard, which covers the physical (PHY) and medium access control (MAC) layers for low-power and low-bit-rate communications.

The integration of low-power circuits and wireless modes of transmission have made tiny sensor nodes effective for a low-rate and low-power use in wireless sensing applications. WSNs have a wide variety of applications because they can easily adapt to different environments. They can work independently where human presence is hazardous or unthinkable. The arrangement of WSNs has created the opportunity to use numerous smart sensors in different sensor network environments for data collection, processing, analysis, and broadcasting [16]. The WSNs have experienced noticeable development over the previous decade because of their versatility and fast technological improvements [17]. WSN is a fundamental technology used in developing IoT and is viewed as the future extension of the Internet. Sensed data are gathered through the network and transmitted to the sink. An unauthorized user can still abuse WSN due to the quantity of sensors deployed in an environment and the resources to manage the deployed sensors are limited. To avoid manipulation and misuse of the data collected by the sensors, access rights should be granted to only authorize users. However, in the context of IoT, to design an access rights control scheme for WSNs is not easy because the resource of the sensor nodes is very limited [9].

ZigBee is an example of WSN. ZigBee is an application protocol specified in IEEE 802.15.4. The PHY and MAC sublayer were designed for devices that communicate within a short range and are useful in IoT. It provides application-level standards and interoperability among devices talking to other devices with low bandwidth on a long life span battery [18].

2.2.2 Radio Frequency Identification (RFID)

RFID technology is the main revolution for embedded system models mainly for designing and developing microprocessors for wireless data communication. They act as an electronic barcode for automatic item identification [16]. There are two types of RFID tags: active and passive. The active RFID tags have their power source, similar to lower-end nodes of WSN, regarding limited processing capability and storage. The benefit of these tags is that they provide a signal to a reader within the range of 50 m and also have their battery supply capable of delivering instant communication. Active RFID devices have limited life spans [19]. The passive tags do not need batteries but use the power from the reader's inquiry signal to connect the tag to the RFID reader. It can be found in many applications, such as bankcards and road toll tags. Passive Tags are tiny and have virtually unlimited life span.

2.2.3 Wireless Fidelity (WiFi)

WiFi is another wireless communication technology that enables IoT. It gives Internet access to devices within an access point's communication range. It is popular because it is cost-effective, flexible, convenient, and offers constant connectivity.

WiFi is defined in the IEEE 802.11 standard specification, which is now the most used standard. It allows wireless terminals to connect. WiFi is a wireless communication protocol that can connect multiple portable computing devices together not only between themselves but also to the Internet. It is vulnerable to security threats and attacks such as hacking, man-in-the-middle attack (MitM), protocol tunneling and signal lost [20]. Securing WiFi network is an issue that need to be addressed.

2.2.4 The 4G, Long-Term Evolution (LTE), LTE-Advanced (LTE-A) and 5G

The Long-Term Evolution (LTE) standard was defined as a transition technology towards a 4G cellular network. It provides new features that support the scope of high bandwidth and improves network performance and cost-optimized IoT devices. LTE can be divided into two stages, LTE and LTE-Advanced, designed for 4G capacity, spectrum and service requirements. Multiple-input multiple-output (MIMO) and orthogonal frequency division multiplex (OFDM) are the most fundamental techniques adopted in LTE and LTE-A systems. MIMO and OFDM can concurrently meet the scope and data rate requirements for indoor and outdoor communication channels. The features of MIMO include single-user beamforming (SU-BF) and multi user beamforming (MU-BF) [21]. The following are the basic features of LTE-A:

a. In terms of speed, it is almost three times faster than the simple LTE network and consists of carrier accumulation.
b. Also, regarding MIMO, the LTE-A has a well-coordinated and better transmission station for heterogeneous networks than 4G LTE.

The 5G is an enhanced technology that will be introduced by 2020 and is user-oriented. It has a very high speed and massive data bandwidth (up to 10 Gbit/s) with a comparatively low latency and low cost per bit [22, 23].

2.3 Security Issues of IoT Communication Technologies

This section explores security issues that arise from each IoT communication technology.

2.3.1 WSN

Sensors and actuators are the main elements of WSN, which make WSN flexible and incur high latency during communication. These characteristics make WSNs prone to attacks. The security issues of WSN include denial of services (DoS), exhaustion, unfairness, Sybil, flooding, jamming, tampering, and collisions. The characteristics of IoT networks, for example, the exposed nature of wireless communication, dynamic change in network topology, and the constrained capabilities of sensor nodes in terms of processing power, storage, bandwidth usage, and energy, make authenticating the validity of public key certificates heavyweight for sensor nodes [24]. Like other networks, security issues in WSN can be summarized into data confidentiality, authenticity, integrity, and freshness. These four security issues can be solved using cryptographic algorithms, key management, secure routing, and node trust [25].

In WSN, data confidentiality and integrity during communication are the fundamental parts where cryptographic security are mostly needed, which can be achieved through encryption [26]. A cryptographic algorithm is an essential approach to guarantee network security at the PHY layer of IoT. It is also the basic technique for guaranteeing the security of the IoT network layer. Data encryption algorithms are categorized into Symmetric Encryption Algorithms (SEA) and Public Key Encryption Algorithms (PKE). SEA is broadly utilized as a part of WSN due to its simplicity and a small amount of computation, which is compatible with WSN, which has limited storage. The drawbacks of SEA are poor scalability, key confidentiality and authentication [27]. Also, when authenticating using SEA, the authentication code increases computational load and incurs additional power consumption.

In view of these drawbacks, researchers such as Bohge and Trappe [28], Zhu et al. [29], and Jing et al. [30] started considering the application of the PKE algorithm to WSNs. The advantage of PKE over SEA is that every node holds its own private key during communication while the BS holds the public key. The PKE algorithm scales well and does not complicate the key management protocol. It is suitable for node verification and also guarantees the security of the entire network [31]. Presently, three PKE algorithms are suitable for WSN, namely Rabin's Scheme, NtruEncrypt, and ECC. These algorithms have been approved on the Mica2 series wireless sensor platform. Owing to the constrained energy and computing power of WSNs, PKE can be used in securing nodes when the appropriate parameters for optimization are selected [28, 29]. In WSNs, key management stands as the basic security challenge that needs further research because how to develop a lightweight secret key distribution solution that will incorporate all security services, all protocols and application for constrained sensor devices is the major problem in this field.

Due to the latest improvement in sensor hardware and the optimization of software [30], key distribution based on public-key algorithm has gained attention. Research in public-key algorithm are classified into two, namely public key distribution issues and allocation of additional keys using public key algorithms. There are many challenges and drawbacks when using public key certificate distribution in the conventional schemes in sensor networks. Therefore, Jing et al. [30] proposed truncated certificate, which is an implicit solution on how to manage certificate in order to reduce energy consumption.

Karlof and Wagner [32] explored public key and symmetric key to take advantage of their strengths to produce a key distribution scheme. In summary, present research direction on key management focuses on how to reduce complexity in key distribution frameworks. Current researchers use the symmetric key algorithm to design key management for WSNs, which has the advantage of reducing power consumption on sensors and makes it easy to use. Secure routing protocols play an important role in WSN because an attack on the routing protocol directly breaks down the entire network. Subsequently, there is a major focus on how to set up a secure and an effective routing protocol [32]. Due to the power constrain, computing capability and storage limit, routing protocols of conventional network cannot be used in WSN. Even ad hoc network routing protocols usually give rise to new problem in WSNs.

The development of a secure routing protocol plays an important role during authentication of routing data. The authentication process in WSN involves incorporating certification features between nodes, which is quite different from the conventional E2E authentication scheme that uses SSH and SSL. Presently, research in this area is classified into two:

a. Secure routing protocols developed precisely for WSN [33]. An authentication protocol and security verification of constrained devices should be discussed in detail, and the energy management issues in WSN should be considered to achieve a good result in practical applications.

b. Analysis of potential vulnerabilities of routing protocols. Karlof and Wagner [32] discussed the security issues in secure routing and countermeasures attacks in WSNs. Challenges such as DoS and sinkhole are evaluated and solutions proposed based on different attacks.

In WSNs, trust management of nodes have the following features and challenges: resource-constrained nodes, easy nodes capturing and a unique mode of communication (i.e., sensor nodes collect data and report to the BS). These features and challenges expose sensor network to different threats and attacks. In this regard, to secure WSNs, there is a need to introduce a trust management mechanism for sensor network because security cannot be guaranteed when password mechanisms and cryptographic algorithm are used in sensor network.

The purpose of trust management is to solve security issues related to service of open network. To this point, research in the domain of trust management is divided into four: (1) mechanism for measuring trust, (2) mechanism for evaluating trust, (3) mechanism for formalizing trust relationship, and (4) mechanism for formal derivation of trust. Trust update is another essential aspect in trust management that needs to be considered, because trust update is even more critical in dynamic networks, such as IoT. Saha et al. [34] examined how to measure the degree of trust based on communication between nodes.

Due to the characteristics of WSNs, trust management must be given a detailed consideration. For instance, energy consumption on nodes is a serious issue in WSN. Trust management framework must have the ability to strike a balance between the size of the devices and network security. Al-Turjman et al. [35] proposed a trust management security framework for sensor networks. They discussed trust management issues in WSNs and proposed some solutions, such as lightweight cryptographic algorithms, key management and secure routing protocols for nodes. A good lightweight cryptographic algorithm is needed to satisfy the high data security requirement. Key management is a vital issue in WSNs because it serves as the root of security for other IoT components. To prevent WSNs from collapsing, secure and effective security protocols need to be set up, because attackers always directly attack the routing protocols. Finally, an effective trust management solution to secure the network is required to ensure the security of the entire WSNs. In addition, ZigBee is also prone to some security issues and challenges, such as packet manipulation, hacking, key exchange, and Killerbee.

2.3.2 *RFID*

The possible security issues that arise from RFID are tracking, DoS, repudiation, spoofing, alteration, corruption and deletion, eavesdropping and counterfeiting. In the context of IoT, RFID technology is mainly used as for automated exchange of information without any manual involvement and can work in harsh environments [30]. While RFID is broadly used, it faces many security issues.

Below are the three typical security issues of RFID:

a. Uniform coding: Presently, Universal Identification (UID) is the common standard used for RFID tags. A universal standard has not yet been designed [36], but UID standards are supported by Japan, Europe and Electronic Product Code (EPC). A lack of universal standard creates a challenge for readers to access tag information when errors occur during reading process.
b. Conflict collision: Conflicts occur when multiple RFID tags communicate with the reader simultaneously, which makes the reader to not read the data accurately [37]. Implementing an anti-collision mechanism can prevent different tags from transmitting data to the reader concurrently.
c. Trust management: In IoT, node security is more important. Hence, the IoT RFID framework requires a trust management scheme. Trust management is meant for communication between readers and tags and between the readers and the BS [38].

For this purpose, the use of digital signature technology (DST) is very important for IoT RFID framework.

2.3.3 WiFi

Radio communication, transmitters and receivers are the main components of WiFi, which enhance guest access, increased mobility, allow network expansion and supports collaboration. WiFi is vulnerable to the following threats: rogue access points, malicious phishing, misconfiguration, hacking, signal lost due to DoS, war dialing, protocol tunneling and MitM, and DDos/Dos attacks. The security risks in WiFi are classified into two: one from the network trap and the other from the network attack. Network trapping involves compromising a user's account and password when encountering a phishing site. It occurs when a user accesses the Internet. A network attack is when an attacker captures a node, gains access to critical resources and information, and interferes with other communication nodes.

2.3.4 The 4G LTE, LTE-Advanced (LTE-A) and 5G

LTE provides new features, such as high bandwidth, that support IoT devices in terms of cost-optimized, computational time and low energy consumption. The security issues in 4G LTE, LTE-A and 5G network are explained below.

4G LTE, LTE-A and 5G networks are accompanied with security issues and challenges, such as data exposure, data inadequacy and unauthorized attacks. These issues and challenges can be solved by introducing a suitable security key management mechanism and behavioral entity authentication [7].

A comparative analysis of the 4G LTE, LTE-A and 5G network access and the security issue experienced in the networks shows that the emphasis on the security issues in these networks are different from other networks. In the access network, 4G LTE, LTE-A and 5G need to be more concerned with authorization and its solution [7].

The 4G LTE, LTE-A, and 5G technologies are integrated into portable intelligent terminals, for example, smart phones, tablet PCs, and notebooks. In the core network, these technologies are mainly concerned about the security and privacy issues in data communication, such as data confidentiality and authenticity, and their implementation methods. The security issues include theft at the privacy-aware layer and eavesdropping. At the network layer, DDOS attack is a concern. However, 4G LTE, LTE-A and 5G networks have similar security problems if they are used for accessing the network or as the core network, including DDos/Dos attacks, phishing attacks and identity attacks. Unfortunately, there are no good solutions to solve DDos/Dos attacks in 4G LTE, LTE-A, and 5G networks. Still, identity and phishing attacks can be solved using behavioral entity authentication and other techniques [39].

2.4 IoT Communication Channel: Challenges

Communication in IoT involves information exchange/sharing among IoT devices or between different IoT layers. With the huge potential of IoT in many domains, the entire communication infrastructure of the IoT is faulty from the security perspective and is vulnerable to loss of confidentiality for the end users. The IoT communication medium is extremely vulnerable to several attacks, and it is easy to attack them physically. The possible attacks are described below.

2.4.1 MitM Attacks

For data integrity to be maintained during a conversation, attacks like MitM must be prevented. In MitM, an attacker quietly transmits and probably modifies the communication between two IoT devices that are directly communicating. Critical reliable information like patient health status, smart grids' billing information or even the house door's secret key can be forged and altered by an adversary, causing serious security problems [40]. MitM attacks represent a genuine threat to IoT security since they allow the attacker to seize control of a communication channel and access sensitive data in real-time. When this happens, the attacker now has control over the channel. The attacker then forms a connection to the actual node and acts as an intermediary to read, redirect, insert, and modify the traffic between the user and the authentic node.

MitM attacks create challenges in protecting data security and privacy. For example, the security problem in IoT involves active interference by intruders on the devices, i.e., allowing unauthorized users to spy on data through the back-door. Lightweight cryptographic protocols are considered to provide communications security for IoT devices over a computer network as part of DTLS. The MitM attack exploits the shortcomings in the communication parties' present authentication protocols [41].

2.4.2 Eavesdropping

Eavesdropping is an electronic attack on the communication channel, i.e., wired or wireless networks, where communications are interrupted by an individual to extract data from the information flow [42]. Eavesdropping becomes really simple when the greater part of the communication is wireless. In IoT, eavesdropping occurs at the network layer and takes the form of data sniffing. A specialized program is used for sniffing and recording packets from the network layer and, subsequently, listened to or read using cryptographic tools for analysis and decryption. Privacy is used to provide efficient access control and security against eavesdropping.

2.4.3 Counterfeiting Attacks

Counterfeiting is simply imitating or forgery. IoT devices require lightweight security. Unfortunately, this makes them vulnerable because an active attacker can easily duplicate and modify the content of the devices [43]. The IoT technologies can be exploited in marketing applications for stock-keeping inventory and product availability. They can likewise assume a part in secondary selling support, whereby customers can naturally recover all information about the items they purchase. Also, identification technologies will provide a unique identifier on items/products, which will help in restricting thefts, fighting counterfeiting, and providing a complete, trustworthy description of the items/products [44].

2.5 Security Issues in IoT Network

This section introduces the security issues in IoT networks.

2.5.1 Architecture Security

Architecture security is a unified security scheme that addresses the requirements and potential risks of a certain situation or environment. It also specifies when and where to apply security controls. The design principles are reported clearly in security architecture, and in-depth security control specifications are generally documented. However, presently, in IoT, there is no unified security architecture [45]. Security is a significant challenge that needs proper attention for IoT to be fully adopted. IoT architecture is expected to manage a projected number of billions of interconnected things. This will create an amazing number of paths that are accessible by malicious attackers because global availability and connectivity are the basic visions of IoT. At the moment, there is no universally accepted IoT architecture [46], which makes IoT devices vulnerable to attacks and threats, although several researches, such as Chakrabarty et al. [18], Valdivieso Caraguay et al. [47], Akhunzada et al. [48] and Ramão Tiago Tiburski et al. [49], etc., have focused on IoT architecture. Specific existing solutions for IoT architecture for different application domains are discussed in this section, and a summary is tabulated in Table 2.1.

A secure IoT architecture for smart cities that addresses the vulnerabilities in traditional IoT systems was proposed by Chakrabarty et al. [18]. The architecture

Table 2.1 IoT security architectures and application domain

Refs.	Architecture	Application domain	Objectives
[47]	SDN architecture	Smart environment	To eliminate the rigidity present in traditional networks
[50]	Secure and efficient authentication and authorization (SEA) architecture	Healthcare	To develop a secure and efficient verification and authorization framework for IoT-based healthcare systems
[51]	Smart city architecture	Smart city	To facilitate the interaction remote sensor systems and data and communication technologies
[49]	SOA	Smart transportation	To define a secure middleware architecture IoT services To analyze and make deliberations on security services that can be used IoT middleware
[52]	Conceptual organizations framework	Business organizations	To identify two types of IoT architecture for an organization: cloud-centric three-layered and autonomic oriented five-layered architecture
[18]	Black SDN architecture	Smart city	To address the vulnerabilities in traditional IoT systems

comprises Black Networks and a Key Management System (KMS) that provides confidentiality, integrity, and privacy. It aims to deliver security services that will mitigate the vulnerabilities in IoT networks, specifically for mission-critical data, at the Link and Network layers. However, this approach's drawback is that it did not create a privacy solution in defining the location of things. In addition, header encryption makes a new routing challenge for asleep IoT nodes, which leads to data loss. Caraguay et al. in [47] and Akhunzada et al. in [48] introduced an SDN architecture for developing IoT applications to eliminate the inflexible security nature present in traditional networks. According to the authors, the purpose of adopting an SDN architecture is to provide a basis for developing a secure Network Operating System (NOS) that allows administrators to have a global view of possible threats and attacks on an IoT network and give them the privilege to control the network against the threats. Nevertheless, security, scalability and reliability of connected things are some of the drawbacks in SDN. Furthermore, the separation between the control plane and data plane in SDN mostly causes poor performance in packet processing, which in turn leads to significant problems, such as packet delay or loss and DDoS [53].

Similarly, a novel SDN-based security architecture for IoT, also known as SDN domain using Border Controllers, was proposed by Flauzac et al. [54]. The authors describe how SDN can be used to interconnect heterogeneous IoT, how the security of each domain can be enhanced, and finally, how the security rules are distributed without compromising the security of any domain. However, how to secure both wanted and unwanted traffic and enterprise protection are the major drawbacks of using Border Controllers, and the authors do not address these challenges.

Tiago et al. in [49] discussed the advantages of having a well-defined standard security architecture for service-oriented architecture (SOA)-based IoT middleware. They studied the approaches taken by different researchers. They outlined the security facilities that can be used when defining IoT security architecture to reduce the security threats in SOA-based IoT middleware frameworks. SOA-based methods also provide IoT applications with an inflexible and organized reflection of security facilities needed for communication by things (IoT devices). It helps ensure high levels of system interoperability, providing system services based on devices and applications. Nevertheless, the coexistence of SOA and Resource-Oriented Architecture (ROA) creates a new set of traditional security demands that must be followed for resource-constrained environments to guarantee system safety, and none of the analyzed works suggests solutions that spell out all the middleware security requirements. So, a security countermeasure system in the middleware architecture is required to protect the IoT middleware from attacks.

Moosavi et al. [50] proposed a distributed smart e-health gateway architecture for IoT-based healthcare systems. The architecture depends on a certificate-based DTLS handshake protocol, the elementary IP security key for IoT. The architecture can adapt to security challenges in general healthcare systems, such as scalability, trust, and consistency. Node authentication is the main limitation in this architecture because sensors used in medical applications are extremely resource-constrained and cannot adapt to cryptography methods requiring heavy computations.

Vučinić et al. [45] proposed an object security architecture (OSCAR) for E2E security in IoT. OSCAR was evaluated in two ways: using 802.15.4 LLN and M2M communication in two hardware stages and using MAC layers on a real testbed alongside the Cooja emulator. The architecture utilizes authorization servers to grant access to users, permitting users to demand resources from the Constrained Application Protocol (CoAP) nodes. OSCAR supports multicasting that mutually handles the issues of E2E security and authorization. The latency of the Elliptic Curve Digital Signature Algorithm (ECDSA) authorization is the drawback of the architecture, which greatly affects the microcontroller unit (MCU) and computation capabilities of IoT devices.

Xi et al. [55] proposed an appropriated secure service structure approach with data stream control in service clouds. The approach benefits cloud-based architecture by providing users with a new form of constituting elementary services to accomplish complex tasks. In a dynamic structure that involves different types of services, the approach first authenticates every service component by model checking and later, guarantees the data stream security amid the procedure of service structure. In various cloud environments, outsourcing information and applications is an important task for data stream security because sensitive data may be leaked to some unapproved users/attackers during service confirmation.

The absence of a unified IoT architecture makes IoT communication technologies such as WSN, RFID, WiFi 4G LTE, LTE-A and 5G prone to different attacks. It also makes the data in the communication channel vulnerable to threats like eavesdropping, MitM and counterfeiting attacks.

2.5.1.1 Existing IoT Framework

An IoT security framework comprises architectures that manage and provide access control, authentication, and authorization to protected devices. In this section, some of the existing authentication frameworks are discussed, enabling us to know the various types of security frameworks for resource-constrained devices.

a. **Identity Certificate-Based Framework**

The Identity Certificate-Based Framework is based on identity certificates that permit users to authenticate each other without prior connection, and the messages are encrypted and digitally signed. The function of the certificates is to bind a public key to an identity. The Public Key Infrastructures (PKIs) and Pretty Good Privacy (PGP) are typical examples of this type of framework [56].

b. **Single Sign-On (SSO) Framework**

The SSO is a non-multiple password framework that grants users authentication once, after which users can access all the system resources [57]. The SSO framework is classified into Kerberos, Windows Live ID, OpenID, Liberty Alliance, and WS-Federation. The Kerberos is a distributed authentication framework that provides SSO inside a single administrative domain [58]. The windows live ID is an Internet-based SSO framework commonly used by Microsoft applications and web services, for example, MSN Messenger [59]. OpenID is an authentication framework with a single digital identity that enables users to use a single username and password to login into various web sites, eliminating the need for distinctive usernames and passwords for each site [60]. The liberty alliance is a consortium that aims to provide open standards, procedures, and preeminent practices for federated identity management. The WS-Federation is a federated identity standard that forms some part of the web services security framework [61].

c. **Federated Identity-Based Framework**

The federated identity-based authentication framework enables users from one security domain to access the resources in another without requiring another user's account. It is based on trust between both domains. In this case, users will register their credentials in their domain with the authentication server, while the other domains trust its assertions.

d. **User-Centric Identity Management Framework**

The user-centric identity management framework is developed from the user's perspective, focuses on device usability, and is cost-effective. An approach such as information cards, which give users a single identity and give users control over access to their resources, are used in user-centric identity management for handling multiple identities.

e. **Device Security Framework**

The device security framework comprises security capabilities and device resident security software distributed across the network. They are embedded in resource-constrained devices to provide security at the physical level during the manufacturing stage.

2.5.2 Authentication

Authentication is a verification process of a device's identity to ensure the authenticity of the device. Authentication processes are granted to devices that possess access control. With the nature and threats imposed on IoT networks, there is a need to introduce secure identity and access control on devices [62]. Devices like sensor nodes or RFID do not have access control functions and exchange information freely with each other. For instance, RFID communication between a tag and a reader can easily lead to security issues, where an attacker can access the tag and obtain the result through eavesdropping [63]. So, there is a need to establish an authentication and authorization scheme among devices to attain the security goals for IoT networks.

Authentication includes confirmation between routing peers of connected IoT devices before exchanging the route information (known as peer authentication) and guaranteeing that the source of the route data is from the connected peer devices (known as data origin authentication). This helps enhance the primary element in IoT vision, M2M communication [64]. The motivation behind the authentication protocol is to provide access to authentic IoT users. In the IoT application domain, authentication allows the integration of different IoT devices and their deployment in various smart environments, such as smart cities. A smart environment can merge different services provided by different multiple shareholders and scales to support numerous users in a dependable and distributed way.

Weber in [65] and Weber et al. in [66] introduced a new identity and access management approach for the impending Internet comprising The authors' emphasis on E2E secure communication and user privacy issues. Their result is built on utilizing a trusted personal device called Minimal Entity. However, for this approach to guarantee user identity, it needs the Trusted Platform Modules (TPM) alongside the minimal entity. Brachmann et al. in [67] worked on E2E transport layer security in IP-based IoT that uses resource-constrained devices and high-performance machines. This approach aims to guarantee E2E security among two IoT devices through the mapping of TLS and DTLS. However, TLS/DTLS in a LLN with a back-end service that does not support CoAP. Bonetto et al. in [68] proposed a secure E2E communication approach between IoT devices and ICT devices. Trusted unconstrained devices that are used for offloading heavy computational operations were introduced. However, the solution did not create node authentication with key establishment mechanisms or as a users privacy solution. Instead, it concentrates mainly on securing communication between IoT and ICT devices.

Gaur et al. in [51] proposed ID authentication at IoT sensor nodes. The approach is based on a one-time cipher request-reply scheme. The scheme uses a pre-shared matrix with a dynamic variable cipher when communication involves multiple parties. The communicating parties create a random coordinate as the key (i.e., password) coordinate. All messages among parties are encrypted using a key, node ID, and timestamp. The communicating parties authenticate the timestamps and can also use the timestamp to cancel a session. However, this approach is only efficient in an IoT domain where securing things is not exceptionally delicate and significant because the key can be rehashed for various coordinates. If the password is changed consistently, security can be enhanced for that specific IoT framework. In addition, establishing a pre-shared matrix needs to be secure for this work to be implemented for many IoT devices.

Gubbi et al. in [16] focus on a common authentication scheme for IoT between different layers and terminal nodes. The scheme is based on hashing and element extraction. The extracted element is mutually shared with the hash function to dodge jamming attacks. This scheme essentially provides a good security solution for authentication in IoT. The extraction procedure has some irreversibility properties (which is lightweight) that guarantee the security of connected things in IoT domain. The scheme emphasizes the authentication process between different IoT layers that send data to terminal nodes, not the reverse. The claim that the scheme will enhance data security is only theory-based, and there is no practical proof to support it.

Hernández-Ramos et al. [69] focused on lightweight validation and authorization security framework for constrained smart objects. The objects/devices in the proposed security framework comply with the recent IoT Architectural Reference Model (ARM) project presented by EU FP7 IoT-A project. Moreover, the framework subsequently intends to propose a general security method for developing novel lightweight security protocols in IoT. But, the authors did not address the integration of the proposed framework into constrained IoT environments for authentication and authorization and define alternative procedures to evaluate the correctness of this framework.

An identity authentication model for capability based on access control was proposed by Neisse et al. [70] for IoT. A public key technique well-suited for lightweight security approaches, mobile/portable devices, distributed devices, and constrained IoT devices was proposed for use with existing technologies like Bluetooth, 4G, WiMAX, and WiFi. This approach includes a timestamp in the authentication message between communicating devices to prevent MitM attacks. The identity authentication is carried out in three sequential phases.

- Key generation phase: A secret key that is Elliptical Curve Cryptography-Diffie Hellman algorithm (ECCDH) is generated in this phase.
- The establishment phase involves a new device identity after generating the secret key. Identity establishment is done either by one-way authentication or by mutual authentication protocols.
- Implementation phase: This final phase grants access control to authenticated devices to communicate with each other.

Even though this model does not completely prevent DoS attacks, it reduces the risk because resource access is granted to only one ID at a time [71].

Ndibanje et al. [72] proposed a security analysis and improvements in authentication and access control for IoT. Their work primarily breaks down current authentication and access control approaches and offers a practicable protocol for IoT. A simple and efficient secure key establishment based on Elliptic Curve Cryptography (ECC) for the authentication protocol is used to improve device authentication. In addition, a role-based access control (RBAC) was introduced for access control policy on applications associated with IoT networks. Nevertheless, the entire communication procedure for IoT sensor nodes is high. Furthermore, the proposed security protocol has not been practically experimented with [73], introducing an efficient authentication and access control technique based on a general perspective of the security issues for the IoT perception layer. The benefit of this technique is creating a session key based on ECC, which improves mutual authentication between user and sensor nodes. However, this technique only solves authentication issues in the IoT perception layer. Still, the attribute-based access control policy was not addressed due to user and sensor node computation costs.

IoT is an evolving concept involving several technologies, including WSNs, RFID, and cloud facilities. The M2M function is the main building block for the IoT paradigm [74]. Moreover, the IoT paradigm applies in many domains, such as smart city, healthcare, smart grids, and transportation. These devices need to communicate with each other. Every communication needs to be secured in one way or another by protecting and giving users the confidence that their information and communication channels are well secured. However, protecting the IoT is challenging and demanding [75]. Therefore, security and privacy issues must be addressed for IoT to be deployed on a large scale in various domains [76].

2.5.2.1 Existing Authentication Schemes for IoT

This section explores existing authentication schemes in IoT. Figure 2.4 provides the classification of existing authentication schemes for IoT.

Generally, the existing authentication schemes are classified into three [77], as shown in Fig. 2.4.

a. **Password-Based Authentication Technique (Weak Authentication)**: The Password-Based authentication technique provides weak authentication and is prone to attacks.
b. **Challenge-Response Authentication Technique (Strong Authentication)**: This technique is broadly used because its cryptographic features, such as symmetric and asymmetric, offer strong authentication. The major drawback of asymmetric authentication is that it is time-consuming, and its implementation in hardware is expensive.

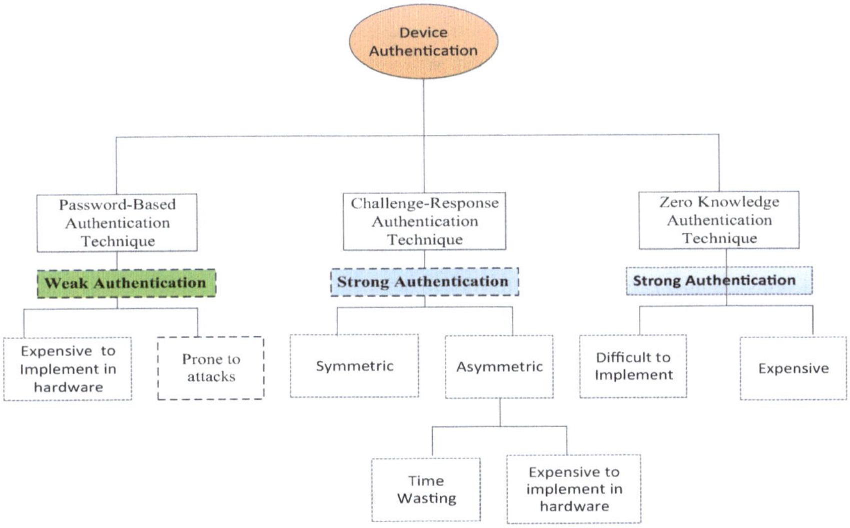

Fig. 2.4 Classification of existing authentication schemes for IoT

c. **Zero Knowledge Authentication Technique**: This technique involves "robust" mathematical problems. It also offers a strong level of authentication, but this technique's mathematical nature makes its implementation very difficult and costly [78].

The challenges imposed on the present authentication techniques discussed make IoT communication technologies (WSN, RFID WiFi and 4G LTE, LTE-A, and 5G) prone to different attacks, such as authenticity of devices, data confidentiality among connected devices, data integrity, and freshness. These challenges also affect the communication channel, making data vulnerable to threats and attacks such as eavesdrops, MitM, and counterfeiting attacks.

2.5.3 *Heterogeneity*

The billions of connected devices have made IoT heterogeneous, making it more vulnerable to threats because each device has different security measures. Due to the discrepancy in the resources of the devices/objects on the smart grid (memory, computation, bandwidth, energy autonomy, time-sensitivity, etc.), and their implemented protocols and communication stacks (for non-IP-based devices), achieving secure E2E communications is a challenging task, often requiring an adaptation of existing solutions or even use of gateways [79]. Resource estimation and management using fog computing for a customer's probabilistic resource estimation (PRE) model was introduced by Aazam et al. [80] to help manage well-organized, successful, and reasonable resources for IoT. Nevertheless, with heterogeneous devices being part of IoT, it is hard to estimate the amount of resources each node will consume, and whether the requesting nodes or devices will completely use the resources they have requested. Also, achieving minimum latency is difficult with devices like healthcare and emergency services because of the unreliable core network that reaches the cloud through shared resources. So, to test for minimum latency, there is a need to apply the model in another research field, such as smart cities, medical centers and smart homes. Moreover, Dunlap and Arce [81] discussed the analysis of available solutions identified with security (i.e., reliability, secrecy, and verification), privacy, and trust in IoT arena. It analyzed the most important accessible solutions identified with security, privacy and trust. The solutions provided by the authors does not well-define the privacy policies that will manage the adaptability of IoT devices in heterogeneous environment.

Persson and Angelsmark [82] presented a framework named Calvin that combines IoT and cloud with a unified programming model. This framework aims to develop a solution that would allow developers to embrace heterogeneity in IoT, but by hiding the protocol and data transport details and improving communication efficiency by avoiding a direct device-to-cloud client/server approach. Calvin is still in its initial phases of development, and presently, no implementation anticipates all security and routing properties required to make autonomous migration for an IoT-distributed environment.

Li et al. [24] proposed a practical access control for sensor networks in the context of IoT. The senders in this novel Heterogeneous Sign-Cryption belong to the Certificate-Less Cryptography (CLC) environment, whereas the receivers belong to the Identity-Based Cryptography (IBC) environment. The main characteristic of this approach is heterogeneity. In particular, the senders and receivers belong to two different cryptographic environments. It permits a sender in the CLC environment to transmit a message to a receiver in the IBC environment. Furthermore, this approach has ciphertext authenticity that allows the shift of the computational cost of the sensor nodes to the gateway. CLC does not require the utilization of certificates. However, it still requires a trusted third party called the Key Generating Center, which generates a partial private key that utilizes the user's identity and a master key. They also focused mainly on the sensor node's computational cost and energy consumption.

The heterogeneity in IoT devices, platforms, communication technologies, and channels makes IoT more susceptible to threats because each device is designed with different technology and security measures.

2.5.4 Quality of Service (QoS)

QoS design is essential for routing data in resource-constrained devices to enable distinguished delivery and to guarantee QoS. Some solutions are provided to improve the services in constrained nodes to ensure good QoS for constrained devices. The solutions include adaptive edge (fog) computing solutions based on regressive admission control (REAC) and fuzzy weighted queueing (FWQ) with adaptive computing methods for IoT networking at the network edges proposed by Jutila [83], which is used to optimize and control traffic flows and network resources. The FWQ control with a feedback mechanism provides properties related to system stability, short settling times, and fast response time. REAC helps in managing the E2E network performance. However, the solutions focused on only one QoS metric, network capacity, and do not address other QoS issues, such as connectivity, reliability, and delay. To avoid network congestion, the operating capacity (i.e., IEEE 802.11p) must fully support two Road Side Unit (RSU) deployments. Unfortunately, it only supports one RSU deployment, which leads to network congestion. Therefore, solutions that address interoperability challenges and the unsolved QoS metrics issues, such as connectivity, reliability and delay, are needed.

In addition, Chakrabarty et al. [18] proposed Black SDN to enhance secure communications by encrypting the header and the payload at the network layer to mitigate a range of attacks and improve IoT networks' overall lifespan and network performance. Resource-constrained IoT nodes cannot support a full SDN implementation and do not address how to secure the black link layer frame. The black link layer frame is defined as an application-delivery network that provides key services for securing data during communication in the network. Therefore, what is needed is a sleep synchronization protocol that is appropriate for black networks to ensure packet delivery to all nodes and secure the black link layer frame by multiple methods that would allow for a fine-grained approach to securing the meta-data, such as (1) replacing the meta-data fields by Grain-128a IV and a keystream, (2) the AES-EAX mode, and (3) a pre-shared IV to allow for better payload efficiency. Homg et al. [84] proposed an adaptive bandwidth allocation algorithm called Adaptive Weighted Fair Queue (AWFQ) for reservation protocols to support QoS in the IoT network layer. The proposed algorithm employs the queue status and priority assignment to control bandwidth sharing among different Internet services and to guarantee that policy for QoS is achieved for resource-constrained devices. The algorithm mainly focused on bandwidth utilization, i.e., how network bandwidth is effectively and efficiently used among resource-constrained devices in a flexible, fair, and prioritized manner. However, bandwidth starvation improvement on resource-constrained devices with low priority and queuing congestion was not addressed. QoS aims to guarantee quality of service delivery among connected IoT devices. Communication technologies like RFID and ZigBee are prone to threats and vulnerabilities such as tracking, DoS, repudiation, spoofing, alteration, corruption, and deletion, exposing data in the communication channel to attacks like eavesdropping and counterfeiting.

2.5.5 Trust Management (TM)

In IoT, node and user privacy are important and must be seriously considered when developing IoT devices. TM involves preserving user privacy, such as user personal data, following the policy and prospect of IoT users in a flexible way. So, in IoT RFID devices, TM is needed to integrate RFID devices. Moreover, TM occurs between readers and the base stations (BS) when communicating, not only between readers and RFID tags alone. In the TM domain, digital signature (DS) technology is employed, and it is very important because it is used for authentication (both on IoT devices and data) and during data communication among different IoT applications [30]. Unfortunately, few researches on TM have been carried out in the IoT domain. The TM aims to solve issues related to security in a distributed environment [38]. Trust is a dynamic concept that can safeguard the existing IoT architecture to provide a uniform decision-making scheme for IoT heterogeneous environments or different domains. Hence, Josang et al. [85] considered TM as the possible solution to security-related issues in IoT. In heterogeneous IoT, addressing trust is demanding because different network nodes have other trust criteria. TM provides an effective approach to assessing trust relationships between IoT devices and helps users be careful when communicating and cooperating with decisions.

Liu and Wang [86] and Yan et al. [4] concentrated on the technologies for controlling heterogeneous connected devices in IoT. The main focus is the heterogeneous network model, trust directing, and TM technology. Their exploration provides direction and strategy for developing future IoT devices. However, there is a need to implement real-life solutions on TM in the IoT domain. Chen et al. [87] explained the complexity of trust relationships among heterogeneous entities. They analyzed the security challenges and threats imposed on IoT, and based on some practicable trust-based ideas gathered, and they proposed a secure IoT architecture. In contrast, Yan et al. [4] only provided some theoretical non-practicable ideas for handling trust in IoT, because the techniques illustrated in their approach (theoretical approach) are unsuitable for resource-constrained devices. Bahtiyar and Ufuk Çağlayan [88] introduced a trust model that focused on extracting trust and providing formal security policies for IoT device/entity when needed. The aim is to provide a formal security policy for an entity on extracting trust data from a secured system for service. Nevertheless, no specific network architecture is considered in this model so that authentication of the parameters used can be properly evaluated and to determine how it can be applied in IoT.

Furthermore, Sicari et al. [89] discussed an analysis of obtainable solutions identified with security (i.e., reliability, secrecy, and confirmation), privacy, and trust in the IoT arena. Trust relationship among two devices supports communications between these devices in future. The devices can always share resources as long as they trust each other. They did not address the implementation of trust negotiation tool that can handle data stream, access control and unified vision concerning the assurance of security and privacy requirements in such heterogeneous environments, which involve various technologies and communication criteria. Therefore, well-defined privacy policies that deal with the scalability and adaptable infrastructure that will manage security threats in a dynamic IoT environment should be developed.

Many studies have recently been conducted on TM for the IoT, and different trust models have been proposed [90, 91]. These trust models may be included in the TM development for the IoT. No related work establishing a trust mechanism has been reported and it remains an open issue for the IoT.

The main objective of TM is to safeguard nodes and users privacy, but in IoT, no TM is established. A lack of established TM will conflict with the security protocols of heterogeneous IoT devices because each connected device is designed using different technology in each heterogeneous IoT environment. For example, the communication technology used for creating RFID are unique identity, auto identification, and unique identity, and for ZigBee are radio, microcontroller and simple protocol. Also the data communication channel for heterogeneous connected IoT devices is prone to attacks such as MitM, channel hijacking, and eavesdropping.

2.5.6 *Jamming*

Jamming belongs to a class of DoS attacks [92] that causes intentional interference in networks during communication through the use of malicious nodes to block authentic communication between devices. Jamming basically disrupts the transmission of a radio signal to delay the utilization of radio frequencies in network. Jamming is performed either constantly or in a discontinuous way or both [93], which causes harmful damage to the network. Jamming attack regularly occurs in the operational phase. For example, at the link layer, counterfeit packets are sent so that they collide with legitimate packets, causing legitimate packets to be dropped [94]. A jammer can be any special equipment, such as, a transmitter. Jamming occurs as a result of abusing the capabilities of this equipment by an attacker through the use of special equipment [95]. The expanding nature and rapid development of embedded IoT devices and the use of wireless technologies make jamming a major research problem and challenge because of its ability to interfere and block communication in the networks. Jamming attack is the one of the most harmful attacks that obstructs the communication channel by presenting counterfeit packets in a network. Jamming is the major threat to IoT, which is a network comprising of little nodes with constrained energy and computing resources. Therefore, it is exceptionally hard to implement the conventional anti jamming techniques in IoT [66]. The IoT devices are practically used and applied in different application domains, such as smart city, intelligent transportation system, healthcare and so on. Protecting these applications is demanding because each application has different security requirements. Hence, security is the main issue in IoT application. IoT devices are tremendously attacked by the various types of jamming attacks [96].

Securing IoT networks is a challenging task because of the characteristics and features of IoT devices [97]. To prevent IoT application from jamming attacks, a conventional cryptographic security mechanism was introduced in the sensor domain to prevent attacks like network spoofing and packet injection. Despite the introduction of conventional cryptographic security mechanisms, IoT is still susceptible to attacks that target the communication medium. At the physical layer of IoT, there are few researches on jamming, which are discussed below. Zhang et al. in [98] Using a classical probability theory, proposed a secure wireless communication for the IoT collision under eavesdrops. It is a theoretical solution based on wireless communication technologies (i.e., WSN) for securing essential information collection procedures for IoT applications. The solution focused primarily on the IoT physical layer (PHY) in a distributed environment. The solution was introduced to overcome the conventional cryptography security technique that are unsuitable for information collection procedure for IoT sensor node with limited computation capacities. They explored the PHY of IoT security techniques to guarantee the security of information collection procedure for IoT devices in a random wireless channel. However, this technique's experiments or real-world implementations have not been performed.

Babar et al. [5] proposed a novel defensive security architecture for IoT. An activity and sequential modelling (ASM) technique was used to protect against jamming attack in IoT communication channel. ASM is an effective tool for developing a defensive mechanism against jamming attacks. The solution in this proposed technique improved the security in IoT communication channel and increased the application domain of IoT. The authors focused only on developing a countermeasure by considering jamming attack at the PHY of a single IoT environment and there is every possibility of contradicting the security countermeasure of another environment when communicating.

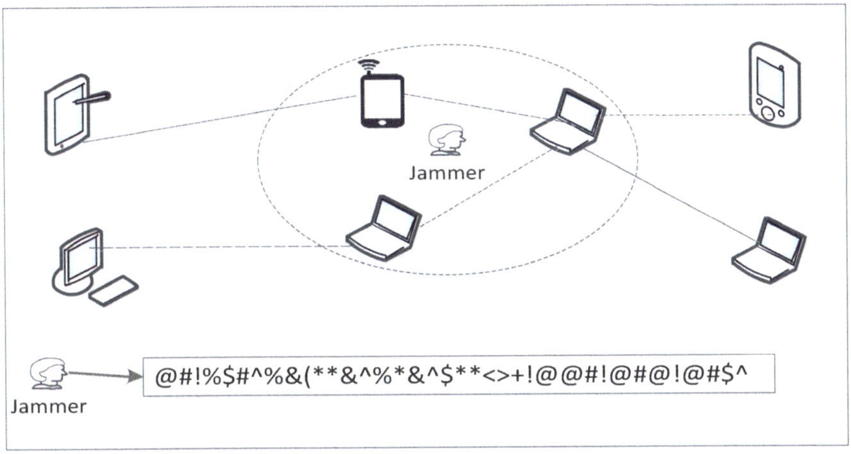

Fig. 2.5 Functional model for constant jammer

2.5.6.1 Existing Jamming Attacks in IoT Networks

The common existing and effective jamming attacks in IoT networks are:

a. **Constant Jammer**

The constant jammer emits radio signals continuously in the communication channel. The signals are emitted randomly and do not follow the MAC protocol underlying principles [5, 99]. Because the MAC protocol enables authentic nodes to send out packets when the communication channel is idle, a constant jammer does not allow the communication channel to be idle. A constant jammer's goal is to continuously emit signals to keep the communication medium busy, thus, preventing legitimate nodes from accessing the communication channel. Figure 2.5 illustrates the functional model of constant jammer.

b. **Deceptive Jammer**

A deceptive jammer continuously sends and inserts random data in the communication channel rather than sending random bits without space/gap between the packet broadcasts. There can be no effective communication between nodes due to constant data streaming in the channel from a deceptive jammer. Figure 2.6 provides the functional view of deceptive jamming in IoT networks.

Here, the deceptive jammer make use of deceptive techniques to inject packets constantly to the communication channel without allowing space between the injected packets, thus, to keep the nodes busy processing packets.

c. **Random Jammer**

The random jamming is an intelligent jamming that conserves energy because it do not make use of unlimited power supply. This is not the case in other kind of jamming,

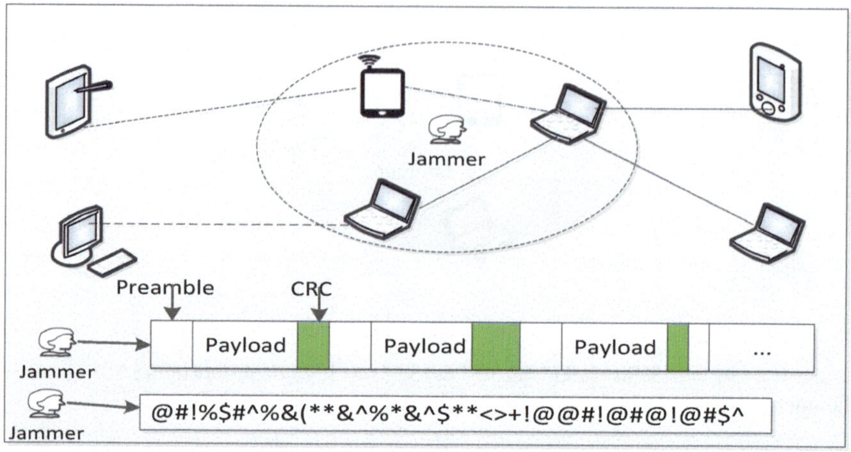

Fig. 2.6 Functional model for deceptive jammer

where a jammer node transmits continuously or after some interval without caring about its own energy level [94]. Here, a jammer node alternates between sleeping and jamming after some regular interval. This kind of jamming behaves like both constant jamming and deceptive jamming. Its unpredictable behavior makes it difficult to detect a jamming attack compared to constant or deceptive jamming. Figure 2.7 presents the functional view of random jamming. However, this jammer is good for jammers that do not have unlimited power supply.

d. **Reactive Jammer**

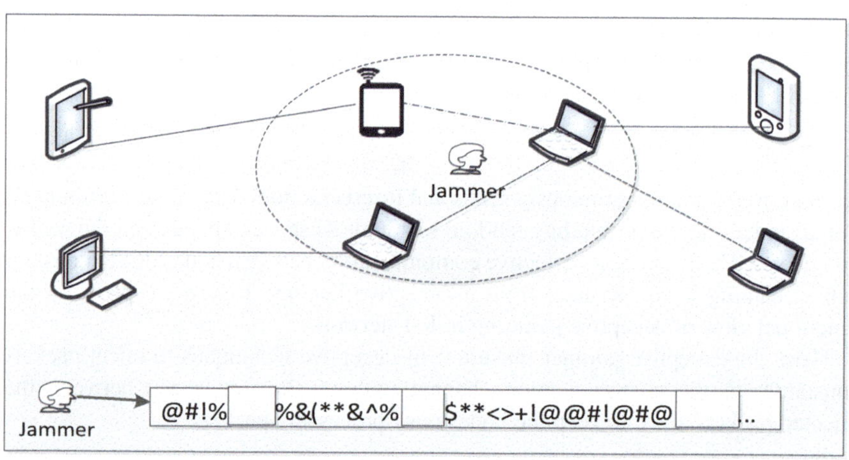

Fig. 2.7 Functional model for random jammer

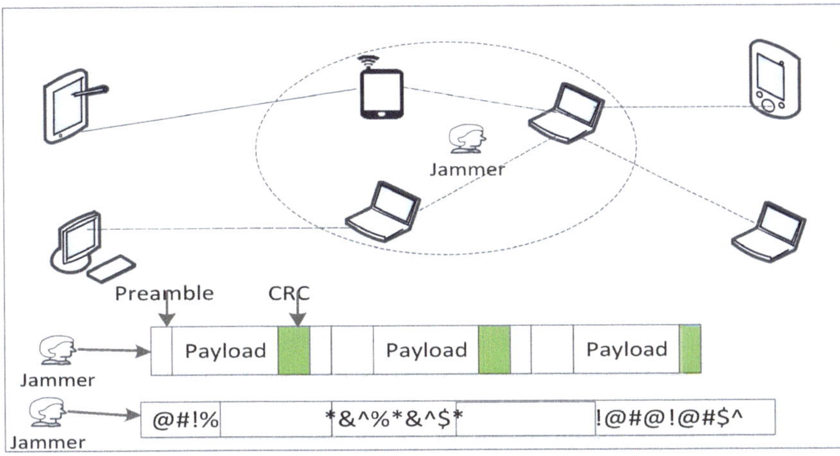

Fig. 2.8 Functional model for reactive jammer

The reactive jamming is the most challenging type of attack to detect. It is more dangerous than previously mentioned attacks in terms of performance. Unlike other jamming attacks, the reactive jammer only transmits signals when it detects that the normal node is sending packets to other nodes; it keeps itself quiet [5]. This makes reactive jamming very difficult to detect and can save energy. The functional view of this type of jammer is illustrated in Fig. 2.8.

The reactive jammer targets the reception of a message and is harder to detect because it transmits when it senses that the communication is active and remains silent when the channel is idle.

2.6 Security Approaches

This section discusses the existing security approaches in IoT. Much work has been conducted into securing IoT devices, but the threshold of minimum resources needed to support these measures leaves many devices lacking sufficient resources to support secure communication.

2.6.1 Lightweight Approaches

Al-Turjaman and Gunay in [100] introduced a lightweight authentication protocol to secure RFID tags. The perception layer of IoT involves devices like RFID and sensors. These devices are constrained in nature and their computational capability is also limited. These characteristics serve as a problematic issue in applying any cryptography algorithms to guarantee IoT network security. When the RFID is insecure, an attacker can easily gain access to the network through sniffing and reprogramming the victim's Electronic Product Code (EPC) tag. This attack can be avoided when an authentication protocol is applied on the tags. The authentication protocol safeguards combine authentication between RFID readers and tagged items with minimum computation overhead on the devices.

Gubby et al. in [16] focused on a common authentication scheme for IoT between different layers and terminal nodes. The scheme is based on hashing and element extraction. The extracted element is mutually shared with the hash function to dodge jamming attacks. This scheme essentially provides good security solution for authentication in IoT. The extraction procedure has some irreversibility properties (which is lightweight) that guarantee security of connected things in IoT domain. The scheme emphasizes the authentication process between different IoT layers that send data to terminal nodes, not the reverse. The authors' claim that the scheme enhances data security is only theoretical presently, and there is no practical proof to support it.

Hernandez-Ramoz et al. in [69] focused on lightweight validation and authorization security framework for constrained smart objects. The objects/devices in the proposed security framework are in compliant with the recent IoT Architectural Reference Model (ARM) project presented by EU FP7 IoT-A project. Moreover, the framework subsequently intends to propose a general security method for developing novel lightweight security protocols in IoT. But, the authors did not address the integration of the proposed framework into constrained IoT environments for authentication and authorization and define alternative procedures to evaluate the correctness of this framework.

Abdmeziem and Tandjaoui in [101] proposed a novel lightweight key management protocol. The protocol depends on the association of different IoT security components to set up a secure and protected communication channel for E2E-constrained nodes and wireless things. The protocol guarantees data confidentiality and constrained node authentication during data transmission along the channel. But E2E secure protocol has some weaknesses. For example, the security protocol is limited to offloading heavyweight cryptographic primitives to unwanted parties and does not specify the needed tradeoff between the communication overhead and the number of third parties.

Yao et al. [102] proposed a lightweight no-pairing Attribute-Based Encryption (ABE) scheme based on ECC to address data security and privacy issues. Their approach reduces the computation and communication overhead in IoT. However, ABE is poor in scalability and inflexible in revoking attribute, which makes it not applicable in multi-authority applications. Therefore, there is a need to develop a lightweight multi-authority oriented ABE and a flexible attribute revoking scheme.

Opoku et al. in [103] proposed a pervasive lightweight verification mechanism for WSNs in distributed IoT applications. The DTLS scheme was used to conduct a security analysis on PAuthKey to measure the security performance of WSNs. The authors implement the PAuthKey protocol and demonstrate its performance capacities on the high resource constrained sensor nodes. However, many security threats and issues, such as access control and multicasting, are encountered with distributed IoT due to network heterogeneity and device mobility. Hence, there is a need to develop an implicit certificate scheme for access control and multicasting on a large scale and implement a security protocol that can handle threats in distributed IoT network applications.

Bose et al. in [104] proposed a lightweight scheme for secure channel establishment to control confidentiality level, evaluate a security score from the fine-grained sensor data, and preserve and protect content over secure transmission. A lightweight security mechanism can support and measure the sensor dataset's confidential value (i.e., affects the secrecy connection) (i.e., data in smart meters). Nevertheless, based on multivariate, the scheme can only consider a single security scenario (i.e., sensitivity) and how to derive sensitivity analysis and privacy degree. It does not address multi-dimensional sensor data. Hence, there is a need to develop an algorithm for deriving sensitivity analysis and privacy measures based on multivariate and multi-dimensional sensor data to extend the scheme to other IoT use cases especially for intelligent transportation.

2.6.2 Model-Based Security Approaches

Neisse et al. in [70] proposed a Model-based Security Toolkit (SecKit) to integrate a model-based security toolkit in the IoT management framework for smart cities that extends the management of trust relationships. SecKit provides a source for security designing tools, add-ons, runtime mechanisms, and extensions that address data security, trust, and confidentiality requirements. Besides, SecKit metamodels can be easily utilized as a theoretical reference for conceptual understandings between various domains in the IoT framework and for improving the interoperability configuration between the various domains. However, there is no addition of trust and risk models no analysis of information flow (both static and dynamic) to specify the causality associated with the behavioral models and runtime events. So, security and privacy solutions should be developed where IoT devices can operate in a dynamic environment. Also, information flow should be implemented to specify the interconnection relationships in models and runtime events for static and dynamic analysis.

Resource estimation and management using fog computing for a customer's probabilistic resource estimation (PRE) was introduced by Aazam et al. [80]. The proposed model was used in IoT for resources management. In addition, it is used for analyzing accessible security solutions identified with privacy and trust in the IoT healthcare domain. The privacy policies in managing the adaptability and heterogeneous setting was not well defined. Also, it is hard to estimate the amount of resources that each node will consume, and whether the requesting nodes or devices are going to completely use the resources they have requested. Finally, achieving minimum latency is difficult with devices like healthcare and emergency services. So, to test for minimum latency, there is a need to apply the model in another research field, such as smart city, medical centers and smart home.

Wu and Wang in [105] proposed a model-based approach for checking and strategizing the implementation policy enforcement in IoT architecture by using standard interfaces, workflow and ontology models. In this model, an enforcement architecture policy was introduced that deals with the fundamental aspects of simulation in IoT environment. In addition, this approach used a model-driven mapping and interpretation to evaluate the prospect cross-domain (heterogeneity) deployment and security policy.

Jararweh et al. in [106] proposed a full software-defined framework model (SDIoT) to improve IoT managing procedures and provide the basic solution for threats in conventional IoT architecture through forwarding, storing, and securing data created from the IoT objects. This is achieved by integrating an SDN, software-defined storage (SDStore), and software-defined security (SDSec) into a single software-defined-based control model. The SDIoT framework accelerates and facilitates the control and management processes of the IoT. It also covers and addresses the difficulties present in traditional architecture. In addition, it enables cloud users to utilize the cloud resources adequately by generating segments/fragments and allowing the transparent flow of information. There are issues regarding SDN compatibility, security, and interoperability, and presently, there is no practical and experimental SDIoT framework for testing different forms of IoT topologies. Therefore, developing an SDIoT framework to investigate various types of IoT topologies that will address security issues and interoperability in SDN is needed.

2.7 Implementation Techniques of Existing Approaches for Preventing Jamming Attacks

This section introduces the security implementation techniques for preventing jamming attacks in IoT. The implementation techniques are classified into software attack and hardware attack implementation techniques.

2.7.1 Software Attack Implementation Techniques for Jamming Attacks

Software attacks are the main source of security threats and vulnerabilities in any framework. Software attacks involve using its communication interface to manipulate the implementation vulnerabilities in any device or system [107]. An example of software attack is intentionally adding malicious codes into the system, leading to virus attacks. In IoT, software attacks are implemented by programming embedded IoT devices' processors. The advantage of this implementation technique is that it is flexible, but security optimization, such as energy consumption and computational capacity, are not considered. This approach is mostly overwhelmed in terms of processing capacity for the embedded processors. However, the solution to this problem is using an optimized General Purpose Processor (GPP) to evaluate the computational capacity of embedded IoT devices.

From the perspective of countermeasures against security attacks, different solutions and implementation techniques for IoT software platforms are provided. An example is a software-level countermeasure solution proposed by Gebotys and White [108] against a side-channel attack. It is a Java-based cryptography countermeasure solution for protecting most cryptographic algorithms. It is a technique that has security support for low-energy embedded devices. An embedded security implementation technique for CPU on dual-virtual systems was introduced by Wilson et al. [109]. The method has good energy efficiency and was developed to prevent side-channel attacks targeting IoT software platforms. High computational time during processing is a major drawback of this technique.

Furthermore, Murphy et al. in [110] proposed an implementation technique using hardware-software public-key cryptography for IoT-Based WSNs. The proposed approach focused on IoT software platforms. It was developed to prevent side-channel attackers. The drawback of this technique is high computational time by IoT devices when communicating and poor energy efficiency for resource-constrained devices. Vaslin et al. [111] implemented a security technique for off-chip memory in embedded microprocessor systems. The method tackled side-channel attacks and other security vulnerabilities in IoT software platforms. Also, this technique's high computational time and energy efficiency consumed by embedded devices during communication remains challenging.

2.7.2 Hardware Attack Implementation Techniques for Jamming Attacks

The rapid developments in computer technology have led to designing portable, faster, and cheaper hardware devices. Subsequently, cryptographic hardware is mostly installed in everything from pay TV units to mobile phones to prepaid cards and access cards (smartcards). Private keys or sensitive data are usually stored in these embedded devices. Hence, compromise of these keys or data may be disastrous, leading to access forgery and loss of privacy. In IoT, a hardware attack is an example of a physical attack that interferes with the hardware components of embedded devices. The cryptography algorithm in the hardware security approach is implemented using Application Specific Integrated Circuits (ASICs), which specifically focus on optimization, such as energy consumption, computational time, and basic security functions.

In the perspective of countermeasures against security attacks, different solutions and implementation techniques for IoT hardware platforms are provided. An example is the hardware cryptography algorithm proposed by Standard et al. [112] for crypto-coprocessors. It was developed mainly for IPSec applications with basic security functions and low energy consumption. A fast and flexible hardware that supports IoT hardware level was implemented by Standard et al. [112] using ECC over various standard prime fields. The technique has a good computational time and protects the hardware platform from side-channel attacks but is poor in energy efficiency. Fons et al. in [113] implemented embedded security for IoT hardware platforms. It is a novel technique developed for personal recognition systems. It is inadequate in terms of computational time and energy efficiency.

2.8 Existing Security Countermeasures for Detecting Jamming in IoT Network

In this section, we provide the existing security countermeasures for jamming attacks. There are many different security countermeasures in this domain with varying defense techniques. But it is impractical to cover all the current countermeasures. In this research, we provide the countermeasures discussed in [114]. We first discuss the types of existing countermeasures and their challenges, the existing "Threshold-Based Jamming Countermeasures (TJC)" and its working algorithm for detecting jamming in IoT networks.

2.8.1 Existing TJC

This section explores the existing TJC network, attacker assumptions, and its working algorithm. TJC is a technique that uses a threshold packet count to detect an attack in the network and cure it.

Existing TJC Network and Attacker Assumptions

This subsection lists TJC assumptions for both a network and an attacker. The assumptions are:

a. The network comprises one BS and n sensor nodes.
b. The capability of each node is the same in terms of equipping.
c. Synchronized clock, propagation model (i.e., two-ray ground) and Omni-directional antenna are used to equip all the nodes.
d. Bidirectional links are used to connect all nodes.
e. All nodes can communicate either via single-hop direct communication or multiple-hop communication.
f. All nodes in the network are randomly distributed.
g. A message is sent to the BS by individual sensor node.
h. The jamming attack can be initiated on any node.

i. Reactive jamming is assumed the attack type and it is initiated once the jammer senses an activity on any node.
j. Both the jamming node and the normal nodes are equipped with the same capability, but the jammer node can generate random messages (i.e., random jamming signal).

2.9 Components for Developing IoT Security Framework

This section discusses the major components required for developing the IoT security framework and the components for designing embedded security. Embedded security simply means incorporating inbuilt security features into an IoT device from the developing stage (i.e., from the start). The following five components are required for developing an embedded security framework for IoT [115, 116]:

a. **Cryptographic Algorithms**

Cryptographic algorithm is needed to secure IoT devices especially during communication in an untrusted network. Cryptographic algorithms are mainly the building blocks for developing a tough security solution for embedded systems. With IoT devices' constrained nature and architectural design, such as limited memory and size, lightweight cryptography are usually deployed inside these devices. ECC lightweight provides the high-level requirements for the devices.

b. **Secure Boot**

Secure boot is a security standard developed by the PC industry to help ensure that systems/devices boots with only trusted software recognized by the industry. The aim of secure boot is to ensure that system trusted state is achieved. Secure boot is based on firmware/Random Access Memory routine (ROM) schedule. To prevent secure boot from attacks, there is a need to provide a module for regular software update.

c. **Secure Storage**

Secure Storage manages access to keys. It ensures that access keys are highly protected from an attacker. An example of secure storage includes on-chip ROM memory and off-chip flash memory. Secure storage involves power cycles and during this process, keys are distributed publicly to both authenticated user and potential attackers. The distributed key can easily be hijacked or cloned by an attacker. Because secret key protection is imperative, cryptography algorithms are needed for key management in IoT framework and a persistent check on secure storage during power cycle.

d. **Secure JTAG**

JTAG an interface mainly used for accessing and debugging sub-blocks of an integrated circuits. It is the basic components for troubleshooting and debugging embedded systems that do not have any other debug-capable communication channel. JTAG can be exploited by attackers by reading the internal memory of registers.

e. **Secure Execution Environment (SEE)**

SEE is a processing unit used for performing application execution securely. Its major components include a secure processor, a secure kernel to provide a communication interface for software and hardware, secure code, and data memory.

2.10 Gap in Research

In this chapter, we have examined the problems in securing data in IoT network and the existing solutions available in solving the problems of jamming attacks. The chapter also presents existing solutions that tackle different security issues in embedded IoT devices. In the context of the gap, we specifically discuss the present security issues related to existing security framework and jamming attacks in IoT networks. Tables 2.2 and 2.3 present the existing work on security frameworks and jamming attacks. The drawbacks and challenges of each work is also stated in the table. The motivation of the research is from this drawbacks and challenges of the existing approaches. For the security frameworks, some of the functions and the features of existing frameworks are provided in Table 2.2.

Table 2.2 compares different security frameworks based on their functions and features for embedded IoT devices. Each framework solves a different aspect of problems in securing IoT devices. As shown in Table 2.3, some researchers have

Table 2.2 Functional evaluation of existing frameworks

Refs.	Framework	Identity certificate management	Single sign-on	Federated identity	User-centric	Device security
[56]	Public key infrastructure (PKI)	Available	N/A	N/A	N/A	N/A
[57]	Pretty good privacy (PGP)	Available	N/A	N/A	N/A	N/A
[58]	Kerberos	N/A	Available	N/A	N/A	N/A
[59]	Windows live ID	N/A	N/A	N/A	N/A	N/A
[60]	OpenID	N/A	Available	N/A	Available	N/A
[117]	Liberty alliance	N/A	Available	Available	Available	N/A
[61]	WS-federation	N/A	Available	Available	N/A	N/A

Table 2.3 Comparison of existing approaches for jamming based on functional optimization and countermeasures attacks

Refs.	Methods	Attack	Throughput	Functional optimization	
				Energy consumption	Traffic delay
[112]	Fast and flexible hardware support for ECC over various standard prime-fields	Side channel	Low	High	High
[110]	Hardware-software public-key cryptography implementations for WSNs	Side channel	Low	High	High
[109]	Embedded security implementation for CPU on dual-virtual-systems	Side channel	Low	High	High
[111]	A secure method for off-chip memory in embedded microprocessor systems	Side channel	Low	High	High
[119]	Hardware-compiler technique toward protecting software platform for embedded systems	Side channel	Low	High	High
[113]	Embedded security: a novel developments for personal recognition systems	Side channel	Low	High	High

proposed framework that support various IoT constrained-devices based on identity certificate management, single sign-on, federated identity, user-centric and device security. Each framework has its own limitation in terms of functional performance. For example, Guneysu and Oder [57] introduced Public Key Infrastructure and Pretty Good Privacy (PGP), respectively, using identity certificate management. However, the framework does not support functions like single sign-on, federated identity, user-centric and device security. Similarly, Peters et al. [118] introduced Kerberos using single sign-on, but the framework does not support other functions like identity certificate management, federated identity, user-centric and device security. In addition, Chehab and Abdallah [117] introduced liberty alliance. The framework has functional features such as single sign-on, federated identity and user-centric, but does not support identity certificate management or device security functions.

For jamming attacks, some of the existing approaches for jamming discussed are provided in Table 2.3.

Table 2.3 compares the existing approaches for jamming based on functional optimization and countermeasure attacks on IoT networks. In IoT, energy consumption, traffic delay, and network throughput are the main parameters to be considered. Although much work has been conducted to secure IoT devices, few are on security frameworks for detecting jamming in IoT networks. Existing solutions address different security elements for data communication in IoT networks, but no single attempt provides a full solution to detect jamming in IoT. All the existing approaches and implementation techniques discussed focused primarily on how to speed up the basic security functions for IoT devices (basic security properties for embedded devices). Also, they do not provide a robust security solution against jamming attack in IoT network. Along this line, there is a need to develop a security framework for embedded IoT devices that will help move security issues from function-centric to system architecture design perspective.

This research focuses on developing a security framework for detecting reactive jamming in IoT networks. Reactive jamming is the most dangerous attack and the biggest threat to IoT network [93], consisting of small nodes with limited energy and computing resources. So, it is very difficult to implement the conventional anti jamming methods in IoT. It obstructs the network communication channel through the introduction of noisy packets. Among all possible security attacks in IoT networks, IoT devices are more susceptible to jamming attacks [93].

However, IoT network demands novel security methods that will help monitor and detect threats. Securing resource-constrained devices remains one of the most challenging issues that needs to be addressed in connected devices. A security framework for IoT must be flexible enough to support the maximum level of security needed by resource-constrained devices (especially embedded devices). The framework should be able to deal with different protocols due to the heterogeneous nature of IoT networks and ensure the adaptability of standard security mechanisms with rapid development of embedded IoT devices. Therefore, this research focuses on the software security perspective of IoT networks. In this research, the main objective is to develop a security framework for detecting and curing reactive jamming attacks in IoT networks.

2.11 Conclusion

The IoT has recently emerged as an important research topic. It integrates different sensors and objects to communicate specifically with each other without human interference. A comprehensive review of the state-of-the-art IoT security threats and vulnerabilities in the context of its application, architecture, communication, and enabling communication technologies are discussed in this chapter. The requirements for the development and large-scale deployment of the IoT are increasing rapidly with major security concerns. However, for the IoT to be accepted and fully deployed, the following security issues and challenges must be addressed.

a. **Architecture Security**

At the moment, there is no universally accepted IoT architecture, making IoT devices vulnerable to attacks and threats. Securing IoT architecture is a significant challenge that needs proper attention in order for IoT to be fully adopted. Since IoT architecture is expected to manage a projected number of billions of interconnected things, this will create an amazing number of paths that are accessible by malicious attackers because global availability and connectivity are the basic visions of the IoT.

b. **Authentication Scheme**

Authentication is a verification process that allows legitimate devices to possess access control. But devices like sensor node and RFID do not have access control function, giving rise to security issues, such as eavesdropping during communication. So, with the nature and threats imposed on IoT network, there is a need to establish an authentication scheme that have some kind of encryption capability that will protect data and the communication channel.

c. **Heterogeneity**

The billions of connected devices has made IoT heterogeneous in nature, and the difference on the resources of the devices and their implemented protocols makes them more vulnerable to threats because each device has different security measure. However, to achieve a secure E2E communications is a challenging task.

d. **Trust Management (TM)**

A TM is a dynamic concept that supports HctI in a reliable way and provides uniform decision-making scheme for IoT heterogeneous environment. In IoT, node and user privacy are very important and must be seriously considered when developing IoT devices. However, trust covers a greater extension than security, therefore, it is complex and hard to build, guarantee and maintain". In heterogeneous IoT, to distribute and compute trust between different networks is a challenging issue because various network nodes have different TM criteria.

e. **Jamming**

Jamming attacks are a subset of denial of service (DoS) attacks in which malicious nodes block legitimate communication between devices by causing intentional interference in IoT networks. Because of the proliferation of wireless technologies, jamming in IoT networks has become a major research problem due to the ease in blocking communication channel.

References

1. Atzori, L., Iera, A., Morabito, G.: The internet of things: a survey. Comput. Netw. **54**(15), 2787–2805 (2010). https://doi.org/10.1016/j.comnet.2010.05.010
2. Fadele, A.A., Othman, M., Hashem, I.A.T., Alotaibi, F.: Internet of things security: a survey. J. Netw. Comput. Appl. **88**(April), 10–28 (2017). https://doi.org/10.1016/j.jnca.2017.04.002
3. Horrow, S., Anjali, S.: Identity management framework for cloud based internet of things. In: Secur.'12 Proceedings of the First International Conference on Security of Internet of Things, pp. 200–203 (2012). https://doi.org/10.1145/2490428.2490456
4. Yan, Z., Zhang, P., Vasilakos, A.V.: A survey on trust management for internet of things. J. Netw. Comput. Appl. **42**, 120–134 (2014). https://doi.org/10.1016/j.jnca.2014.01.014
5. Babar, S., Stango, A., Prasad, N., Sen, J., Prasad, R.: Proposed embedded security framework for internet of things (IoT). In: 2015 2nd International Conferences on Wireless Communication, Vehicular Technology, Information Theory and Aerospace & Electronic Systems Technology Wireless VITAE 2011, pp. 1–5 (2015). https://doi.org/10.1109/WIRELESSVITAE.2011.5940923
6. Botta, A., De Donato, W., Persico, V., Pescapé, A.: Integration of cloud computing and internet of things: a survey. Futur. Gener. Comput. Syst. **56**, 684–700 (2016). https://doi.org/10.1016/j.future.2015.09.021
7. Yinbiao, S., et al.: Internet of things: wireless sensor networks. Int. Electron. Commission (December), 1–78 (2014)
8. Lu, C.: Overview of Security and Privacy Issues in the Internet of Things, pp. 1–11 (2014)
9. Raza, S., Wallgren, L., Voigt, T.: SVELTE: real-time intrusion detection in the internet of things. Ad Hoc Netw. **11**(8), 2661–2674 (2013). https://doi.org/10.1016/j.adhoc.2013.04.014
10. Suo, H., Wan, J., Zou, C., Liu, J.: Security in the internet of things: a review. In: Proceedings—2012 International Conference on Computer Science and Electronics Engineering, ICCSEE 2012, vol. 3, pp. 648–651 (2012). https://doi.org/10.1109/ICCSEE.2012.373
11. Zhou, F., Zhao, B., Tao, M., Bai, X., Chen, B., Sun, G.: A large scene deceptive jamming method for space-borne SAR. IEEE Trans. Geosci. Remote Sens. **51**(8), 4486–4495 (2013). https://doi.org/10.1109/TGRS.2013.2259178
12. Li, L., et al.: CreditCoin: a privacy-preserving blockchain-based incentive announcement network for communications of smart vehicles. IEEE Trans. Intell. Transp. Syst. **19**(7), 2204–2220 (2018). https://doi.org/10.1109/TITS.2017.2777990
13. Milbourn, T.: No Title. https://www.u-blox.com/en/blog/ip-versus-coap-iot-communications (2016). Accessed 15 July 2016
14. Ma, X., Fan, X., Liu, J., Jiang, H., Peng, K.: vLocality: revisiting data locality for MapReduce in virtualized clouds. IEEE Netw (February), 28–35 (2016). https://doi.org/10.1109/MNET.2016.1500133NM
15. Zhu, C., Leung, V.C.M., Shu, L., Ngai, E.C.H.: Green internet of things for smart world. IEEE Access **3**, 2151–2162 (2015). https://doi.org/10.1109/ACCESS.2015.2497312
16. Gubbi, J., Buyya, R., Marusic, S., Palaniswami, M.: Internet of things (IoT): a vision, architectural elements, and future directions. Futur. Gener. Comput. Syst. **29**(7), 1645–1660 (2013). https://doi.org/10.1016/j.future.2013.01.010
17. Porambage, P., Schmitt, C., Kumar, P., Gurtov, A., Ylianttila, M.: Two-phase authentication protocol for wireless sensor networks in distributed IoT applications. In: 2014 IEEE Wireless Communications and Networking Conference, vol. 3, pp. 2728–2733 (2014). https://doi.org/10.1109/WCNC.2014.6952860
18. Chakrabarty, S., Engels, D.W., Member, S.: A Secure IoT Architecture for Smart Cities. (2016)
19. Tsai, C.-W., Lai, C.-F., Vasilakos, A.V.: Future Internet of things: open issues and challenges. Wirel. Netw. **20**(8), 2201–2217 (2014). https://doi.org/10.1007/s11276-014-0731-0
20. Shan, L.: J. Comput. math.pdf **6**(10) (2013)

21. Shanzhi, C., Shaohui, S., Yingmin, W., Guojun, X., Tamrakar, R.: Review Paper: A Comprehensive Survey of TDD-Based Mobile Communication Systems from TD-SCDMA 3G to TD- LTE (A) 4G and 5G Directions, pp. 40–60 (2014)
22. Elzein, N.M., et al.: Managing Big RDF Data in Clouds: Challenges, Opportunities, and Solutions. (2018). https://doi.org/10.1016/j.scs.2018.02.019
23. Hashem, I.A.T., et al.: The role of big data in smart city. Int. J. Inf. Manag. **36**(5), 748–758 (2016). https://doi.org/10.1016/j.ijinfomgt.2016.05.002
24. Li, F., Han, Y., Jin, C.: Practical access control for sensor networks in the context of the internet of things. Comput. Commun. **90** (2016). https://doi.org/10.1016/j.comcom.2016.03.007
25. Li, S., Tryfonas, T., Li, H.: The internet of things: a security point of view. Internet Res. **26**(2), 337–359 (2016). https://doi.org/10.1108/IntR-07-2014-0173
26. Karlof, C., Sastry, N., Wagner, D.: TinySec: a link layer security architecture for wireless sensor networks. Security **3**, 162–175 (2013). https://doi.org/10.1145/1031495.1031515
27. Chen, M., Lai, C.-F., Wang, H.: Mobile multimedia sensor networks: architecture and routing. EURASIP J. Wirel. Commun. Netw. **2011**(1), 159 (2011). https://doi.org/10.1186/1687-1499-2011-159
28. Bohge, M., Trappe, W.: An authentication framework for hierarchical ad hoc sensor networks. In: Proceedings of the 2013 ACM Workshop on Wireless Security—WiSe'13, p. 79 (2013). https://doi.org/10.1145/941322.941324
29. Zhu, S., Setia, S., Jajodia, S.: LEAP: efficient security mechanisms for large-scale distributed sensor networks categories and subject descriptors. ACM Trans. Sens. Netw. **2**(4), 500–528 (2015)
30. Jing, Q., Vasilakos, A.V., Wan, J., Lu, J., Qiu, D.: Security of the internet of things: perspectives and challenges. Wirel. Netw. **20**(8), 2481–2501 (2014). https://doi.org/10.1007/s11276-014-0761-7
31. Palattella, M.R., et al.: Standardized protocol stack for the internet of (important) things. IEEE Commun. Surv. Tutor. **15**(3), 1389–1406 (2013). https://doi.org/10.1109/SURV.2012.111412.00158
32. Karlof, C., Wagner, D.: Secure routing in wireless sensor networks: attacks and countermeasures. In: Sensor Network Protocols and Applications 2013, Proceedings of the First IEEE, 2013 IEEE International Workshop, vol. 1, no. 2–3, pp. 113–127 (2013). https://doi.org/10.1109/SNPA.2003.1203362
33. Wood, A.D., Stankovic, J.A.: Denial of service in sensor networks. Computer (Long. Beach. Calif) **35**(10), 54–62 (2012). https://doi.org/10.1109/MC.2002.1039518
34. Saha, A., To, K., PalChaudhuri, S., Du, S., Johnson, D.: Design and performance of PRAN: a system for physical implementation of ad hoc network routing protocols. IEEE Trans. Mob. Comput. **6**(4), 463–479 (2012). https://doi.org/10.1109/TMC.2007.48
35. Al-Turjman, F.M., Al-Fagih, A.E., Alsalih, W.M., Hassanein, H.S.: A delay-tolerant framework for integrated RSNs in IoT. Comput. Commun. **36**(9), 998–1010 (2013). https://doi.org/10.1016/j.comcom.2012.07.001
36. Welbourne, E., et al.: Building the internet of things using RFID. Internet Comput. IEEE **13**(3), 48–55 (2009). https://doi.org/10.1109/MIC.2009.52
37. Roman, R., Alcaraz, C., Lopez, J., Sklavos, N.: Key management systems for sensor networks in the context of the internet of things. Comput. Electr. Eng. **37**(2), 147–159 (2011). https://doi.org/10.1016/j.compeleceng.2011.01.009
38. Gu, L., Wang, J., Sun, B.: Trust management mechanism for internet of things. China Commun. **11**(2), 148–156 (2014). https://doi.org/10.1109/CC.2014.6821746
39. Frizzo-barker, J., Chow-white, P.A., Mozafari, M., Ha, D.: An empirical study of the rise of big data in business scholarship. Int. J. Inf. Manag. **36**(3), 403–413 (2016). https://doi.org/10.1016/j.ijinfomgt.2016.01.006
40. Devi, R., Kumar, A., Dhawan, V.: A node prioritization based load balancing approach to improve cluster head selection in wireless sensor network. Int. J. Curr. Trends Eng. Res. **3**(2), 8–12 (2017)

41. Mahmood, K., Ashraf Chaudhry, S., Naqvi, H., Shon, T., Farooq Ahmad, H.: A lightweight message authentication scheme for Smart Grid communications in power sector. Comput. Electr. Eng. **52**, 114–124 (2015). https://doi.org/10.1016/j.compeleceng.2016.02.017
42. Pongle, P., Chavan, G.: A survey: attacks on RPL and 6LoWPAN in IoT. In: 2015 International Conference on Pervasive Computing: Adv. Commun. Technol. Appl. Soc. ICPC 2015, no. c, pp. 0–5 (2015). https://doi.org/10.1109/PERVASIVE.2015.7087034
43. Whitmore, A., Agarwal, A., Da Xu, L.: The internet of things? A survey of topics and trends. Inf. Syst. Front. **17**(2), 261–274 (2015). https://doi.org/10.1007/s10796-014-9489-2
44. Miorandi, D., Sicari, S., De Pellegrini, F., Chlamtac, I.: Internet of things: vision, applications and research challenges. Ad Hoc Netw. **10**(7), 1497–1516 (2012). https://doi.org/10.1016/j.adhoc.2012.02.016
45. Vučinić, M., Tourancheau, B., Rousseau, F., Duda, A., Damon, L., Guizzetti, R.: OSCAR: object security architecture for the internet of things. Ad Hoc Netw. **32**, 3–16 (2015). https://doi.org/10.1016/j.adhoc.2014.12.005
46. Jiang, H., Shen, F., Chen, S., Li, K.C., Jeong, Y.S.: A secure and scalable storage system for aggregate data in IoT. Futur. Gener. Comput. Syst. **49**, 133–141 (2015). https://doi.org/10.1016/j.future.2014.11.009
47. Valdivieso Caraguay, Á.L., Benito Peral, A., Barona López, L.I., García Villalba, L.J.: SDN: evolution and opportunities in the development IoT applications. Int. J. Distrib. Sens. Netw. **2014** (2014). https://doi.org/10.1155/2014/735142
48. Akhunzada, A., et al.: Secure and dependable software defined networks. J. Netw. Comput. Appl. **61**, 199–221 (2016). https://doi.org/10.1016/j.jnca.2015.11.012
49. Ramão Tiago Tiburski, F.H., Amaral, L.A., de Matos, E.: The Importance of Being, vol. 44, pp. 95–128 (2015). https://doi.org/10.3732/ajb.0800322
50. Moosavi, S.R., et al.: SEA: a secure and efficient authentication and authorization architecture for IoT-based healthcare using smart gateways. Procedia Comput. Sci. **52**(1), 452–459 (2015). https://doi.org/10.1016/j.procs.2015.05.013
51. Gaur, A., Scotney, B., Parr, G., McClean, S.: Smart city architecture and its applications based on IoT. Procedia Comput. Sci. **52**(1), 1089–1094 (2015). https://doi.org/10.1016/j.procs.2015.05.122
52. Vishvakarma, N.K., James, W., Sharma, R.R.K.: ISSN: 1930-6105 Figure 1: three IoT visions. JIMS **15**(1), 35–43 (2015)
53. Chakrabarty, S., Engels, D.W., Thathapudi, S.: Black SDN for the internet of things. In: Proceedings—2015 IEEE 12th International Conference on Mobile Ad Hoc and Sensor Systems, MASS 2015, pp. 190–198 (2015). https://doi.org/10.1109/MASS.2015.100
54. Flauzac, O., Gonzalez, C., Nolot, F.: New security architecture for IoT network. Procedia Comput. Sci. **52**(1), 1028–1033 (2015). https://doi.org/10.1016/j.procs.2015.05.099
55. Xi, N., Sun, C., Ma, J., Shen, Y.: Secure service composition with information flow control in service clouds. Futur. Gener. Comput. Syst. **49**, 142–148 (2015). https://doi.org/10.1016/j.future.2014.12.009
56. Huston, G., Michaelson, G., Kent, S.: Resource certification—a public key infrastructure for IP addresses and AS's. In: 2014 IEEE Globecom Workshops, Gc Work (2014). https://doi.org/10.1109/GLOCOMW.2009.5360715
57. Guneysu, T., Oder, T.: Towards lightweight identity-based encryption for the post-quantum-secure internet of things. In: 2017 18th International Symposium on Quality Electronic Design, pp. 319–324 (2017). https://doi.org/10.1109/ISQED.2017.7918335
58. Neuman, B.C., Ts'o, T.: Kerberos: an authentication service for computer networks. IEEE Commun. Mag. **32**(September), 33–38 (2014). https://doi.org/10.1109/35.312841
59. Ahmad, Z., Manan, J.A.: 3rd International Conference on Advanced Computer Theory and Engineering (ICACTE), pp. 487–491 (2013)
60. Wang, H., Fan, C., Yang, S., Zou, J., Zhang, X.: A new secure OpenID authentication mechanism using one-time password (OTP). In: 7th International Conferences on Wireless Communications, Networking and Mobile Computing, WiCOM 2016, pp. 1–4 (2016). https://doi.org/10.1109/wicom.2011.6040525

61. Ates, M., Gravier, C., Lardon, J., Fayolle, J., Sauviac, B.: Interoperability between heterogeneous federation architectures: illustration with SAML and WS-federation. In: Proceedings—International Conference on Signal Image Technologies and Internet Based System, SITIS 2007, pp. 1063–1070 (2015). https://doi.org/10.1109/SITIS.2007.148
62. Srinivas, J., Mukhopadhyay, S., Mishra, D.: Secure and efficient user authentication scheme for multi-gateway wireless sensor networks. Ad Hoc Netw. **54**, 147–169 (2017). https://doi.org/10.1016/j.adhoc.2016.11.002
63. Qiu, Y., Ma, M.: A mutual authentication and key establishment scheme for M2M communication in 6LoWPAN. IEEE Trans. Ind. Informatics **12**(6), 2074–2085 (2016)
64. Perera, C., Zaslavsky, A., Christen, P., Georgakopoulos, D.: Context aware computing for the internet of things: a survey. IEEE Commun. Surv. Tutor. **16**(1), 414–454 (2014). https://doi.org/10.1109/SURV.2013.042313.00197
65. Weber, R.H.: Internet of things: privacy issues revisited. Comput. Law Secur. Rev. **31**(5) (2015). https://doi.org/10.1016/j.clsr.2015.07.002
66. S. G. Weber, L. A. Martucci, S. Ries, and M. Max: Towards trustworthy identity and access management for the future internet. In: 4th International Workshop on Trustworthy Internet of People, Things & Services (Trustworthy IoPTS 2010) co-located with Internet Things, vol. 4, no. 2010 (2010)
67. Brachmann, M., Keoh, S.L., Morchon, O.G., Kumar, S.S.: End-to-End Transport Security in the IP-based Internet of Things. (2012)
68. Bonetto, R., Bui, N., Lakkundi, V., Olivereau, A., Serbanati, A., Rossi, M.: Secure communication for smart IoT objects: protocol stacks, use cases and practical examples. In: World of Wireless, Mobile and Multimedia Networks (WoW-MoM), vol. 2012, pp. 1–7. IEEE (2012)
69. Hernández-Ramos, J.L., Moreno, M.V., Bernabé, J.B., Carrillo, D.G., Skarmeta, A.F.: SAFIR: secure access framework for IoT-enabled services on smart buildings. J. Comput. Syst. Sci. **81**(8), 1452–1463 (2015). https://doi.org/10.1016/j.jcss.2014.12.021
70. Neisse, R., Steri, G., Fovino, I.N., Baldini, G.: SecKit: a model-based security toolkit for the internet of things. Comput. Secur. **54**, 60–76 (2015). https://doi.org/10.1016/j.cose.2015.06.002
71. Mahmoud, R., Yousuf, T., Aloul, F., Zualkernan, I.: Internet of things (IoT) security: current status, challenges and prospective measures. In: 2015 10th International Conference for Internet Technology and Secured Transactions, ICITST 2015, pp. 336–341 (2016).https://doi.org/10.1109/ICITST.2015.7412116
72. Ndibanje, B., Lee, H.J., Lee, S.G.: Security analysis and improvements of authentication and access control in the internet of things. Sensors (Basel) **14**(8), 14786–14805 (2014). https://doi.org/10.3390/s140814786
73. Ye, N., Zhu, Y., Wang, R.C., Malekian, R., Lin, Q.M.: An efficient authentication and access control scheme for perception layer of internet of things. Appl. Math. Inf. Sci. **8**(4), 1617–1624 (2014). https://doi.org/10.12785/amis/080416
74. Riad, A.B.M.K., et al.: Plugin-based tool for teaching secure mobile application development. Inf. Syst. Educ. J. **19**(2), 25–34 (2021)
75. Kanuparthi, A., Karri, R., Addepalli, S.: Hardware and embedded security in the context of internet of things. In: Proceedings of the 2013 ACM Workshop on Security, Privacy & Dependability for Cyber Vehicles—CyCAR'13, pp. 61–64 (2013). https://doi.org/10.1145/2517968.2517976
76. Roman, R., Zhou, J., Lopez, J.: On the features and challenges of security and privacy in distributed internet of things. Comput. Netw. **57**(10), 2266–2279 (2013). https://doi.org/10.1016/j.comnet.2012.12.018
77. Gu, Z., Han, G., Zeng, H., Zhao, Q.: Security-aware mapping and scheduling with hardware co-processors for FlexRay-based distributed embedded systems. IEEE Trans. Parallel Distrib. Syst. **27**(10), 3044–3057 (2016)
78. Kim, J.: Secure and efficient management architecture for the internet of things secure and efficient code dissemination. In: SenSys'15 Proceedings of the 13th ACM Conference on Embedded Networked Sensor Systems, pp. 499–500 (2015). https://doi.org/10.1145/2809695.2822522

79. Bekara, C.: Security issues and challenges for the IoT-based smart grid. Procedia Comput. Sci. **34**, 532–537 (2014). https://doi.org/10.1016/j.procs.2014.07.064
80. Aazam, M., St-Hilaire, M., Lung, C.-H., Lambadaris, I.: PRE-Fog: IoT trace based probabilistic resource estimation at Fog. In: 2016 13th IEEE Annual Consumer Communications & Networking Conferences, pp. 12–17 (2016). https://doi.org/10.1109/CCNC.2016.7444724
81. Dunlap, A., Arce, M.C.: 'Murderous energy' in Oaxaca, Mexico: wind factories, territorial struggle and social warfare. J. Peasant Stud. **49**(2), 455–480 (2022). https://doi.org/10.1080/03066150.2020.1862090
82. Persson, P., Angelsmark, O.: Calvin—merging cloud and IoT. Procedia Comput. Sci. **52**(Ant), 210–217 (2015). https://doi.org/10.1016/j.procs.2015.05.059
83. Jutila, M.: An adaptive edge router enabling internet of things. IEEE Internet Things J. **4662**(c), 1–1 (2016). https://doi.org/10.1109/JIOT.2016.2550561
84. Homg, M.-F., Lee, W.-T., Lee, K.-R., Kuo, Y.-H.: An adaptive approach to weighted fair queue with QoS enhanced on IP network. In: TENCON 2011. Proceedings of the IEEE Region 10 International Conference on Electrical and Electronic Technology, vol. 1, no. 01, pp. 181–186 (2011). https://doi.org/10.1109/TENCON.2001.949576
85. Josang, A., Ismail, R., Boyd, C.: A survey of trust and reputation systems for online service provision. Decis. Support Syst. **43**(2), 618–644 (2012). https://doi.org/10.1016/j.dss.2005.05.019
86. Liu, L., Wang, W.: Internet of things: objectives and scientific challenges. J. Comput. Sci. Technol. **26**(6), 919–924 (2010). https://doi.org/10.1007/s11390-011-1189-5
87. Chen, D., Chang, G., Jin, L., Ren, X., Li, J., Li, F.: A novel secure architecture for the Internet of things. In: Proceedings—2011 5th International Conference on Genetic and Evolutionary Computing, ICGEC 2011, pp. 311–314 (2011). https://doi.org/10.1109/ICGEC.2011.77
88. Bahtiyar, Š., Ufuk Çağlayan, M.: Extracting trust information from security system of a service. J. Netw. Comput. Appl. **35**(1), 480–490 (2012). https://doi.org/10.1016/j.jnca.2011.10.002
89. Sicari, S., Rizzardi, A., Miorandi, D., Cappiello, C., Coen-Porisini, A.: A secure and quality-aware prototypical architecture for the internet of things. Inf. Syst. **58**, 43–55 (2016). https://doi.org/10.1016/j.is.2016.02.003
90. Lopez, J., Roman, R., Agudo, I., Fernandez-Gago, C.: Trust management systems for wireless sensor networks: best practices. Comput. Commun. **33**(9), 1086–1093 (2010). https://doi.org/10.1016/j.comcom.2010.02.006
91. Gu, X., Qiu, J., Wang, J.: Research on trust model of sensor nodes in WSNs. Procedia Eng. **29**(909–913) (2012)
92. Kasinathan, P., Pastrone, C., Spirito, M.A., Vinkovits, M.: Denial-of-service detection in 6LoWPAN based internet of things. In: International Conference on Wireless and Mobile Computing, Networking and Communications, no. October, pp. 600–607 (2013). https://doi.org/10.1109/WiMOB.2013.6673419
93. Aman, W., Snekkenes, E.: Managing security trade-offs in the internet of things using adaptive security. In: 2015 10th International Conference for Internet Technology and Secured Transactions, ICITST 2015, pp. 362–368 (2016). https://doi.org/10.1109/ICITST.2015.7412122
94. Raymond, D.R., Marchany, R.C., Brownfield, M.I., Midkiff, S.F.: Effects of denial-of-sleep attacks on wireless sensor network MAC protocols. IEEE Trans. Veh. Technol. **58**(1), 367–380 (2011). https://doi.org/10.1109/TVT.2008.921621
95. Grover, K., Lim, A., Yang, Q.: Jamming and anti-jamming techniques in wireless networks: a survey. Int. J. Ad Hoc Ubiquit. Comput. **17**(4), 197–215 (2018). https://doi.org/10.1504/IJAHUC.2014.066419
96. Babar, S.D., Prasad, N.R., Prasad, R.: Jamming attack: behavioral modelling and analysis. In: 2013 3rd International Conference on Wireless Communications, Vehicular Technology, Information Theory and Aerospace and Electronic Systems, VITAE 2013—Co-located with Global Wireless Summit 2013, pp. 0–4 (2013). https://doi.org/10.1109/VITAE.2013.6617054

97. Mpitziopoulos, A., Gavalas, D., Konstantopoulos, C., Pantziou, G.: A survey on jamming attacks and countermeasures in WSNs. IEEE Commun. Surv. Tutor. **11**(4), 42–56 (2014). https://doi.org/10.1109/SURV.2009.090404
98. Zhang, Y., Shen, Y., Wang, H., Yong, J., Jiang, X.: On secure wireless communications for IoT under eavesdropper collusion. IEEE Trans. Autom. Sci. Eng. **13**(3), 1281–1293 (2016). https://doi.org/10.1109/TASE.2015.2497663
99. Pawar, P.M., Nielsen, R.H., Prasad, N.R.: Behavioural modelling of WSN MAC layer security attacks : a sequential UML approach. J. Cyber Secur. Mobil. 65–82 (2012)
100. Al-turjman, F., Gunay, M.: CAR approach for the internet of things Approche de la CAR pour l ' internet des objets. Can. J. Electr. Comput. Eng. **39**(1), 11–18 (2016). https://doi.org/10.1109/CJECE.2015.2492679
101. Abdmeziem, M.R., Tandjaoui, D.: An end-to-end secure key management protocol for e-health applications. Comput. Electr. Eng. **44**, 184–197 (2015). https://doi.org/10.1016/j.compeleceng.2015.03.030
102. Yao, X., Chen, Z., Tian, Y.: A lightweight attribute-based encryption scheme for the internet of things. Futur. Gener. Comput. Syst. **49**, 104–112 (2014). https://doi.org/10.1016/j.future.2014.10.010
103. Opoku, D.G.J., Perera, S., Osei-Kyei, R., Rashidi, M., Famakinwa, T., Bamdad, K.: Drivers for digital twin adoption in the construction industry: a systematic literature review. Buildings **12**(2) (2022). https://doi.org/10.3390/buildings12020113
104. Bose, T., Bandyopadhyay, S., Ukil, A., Bhattacharyya, A., Pal, A.: Why not keep your personal data secure yet private in IoT?: Our lightweight approach. In: Proceedings of the 2015 IEEE Tenth International Conference on Intelligent Sensors, Sensor Networks and Information Processing, no. April, pp. 1–6 (2015). https://doi.org/10.1109/ISSNIP.2015.7106942
105. Wu, Z., Wang, L.: An innovative simulation environment for cross-domain policy enforcement. Simul. Model. Pract. Theory **19**(7), 1558–1583 (2011). https://doi.org/10.1016/j.simpat.2011.03.005
106. Jararweh, Y., Al-Ayyoub, M., Darabseh, A., Benkhelifa, E., Vouk, M., Rindos, A.: SDIoT: a software defined based internet of things framework. J. Ambient. Intell. Humaniz. Comput. **6**(4), 453–461 (2015). https://doi.org/10.1007/s12652-015-0290-y
107. Mazumdar, B., Ali, S.K.S., Sinanoglu, O., York, N.: A compact implementation of Salsa20 and its power. ACM Trans. Des. Autom. Electron. Syst. **22**(1), 1–26 (2016)
108. Gebotys, C.H., White, B.A.: EM analysis of a wireless Java-based PDA. ACM Trans. Embed. Comput. Syst. **7**(4), 1–28 (2008). https://doi.org/10.1145/1376804.1376812
109. Wilson, P., Frey, A., Mihm, T.: Implementing embedded security on dual-virtual-CPU systems. IEEE Des. Test Comput. **3**(1), 582–591 (2015)
110. Murphy, G., Keeshan, A., Agarwal, R.: Hardware–software implementation of public-key cryptography for wireless sensor networks. In: IEE Irish Signals and Systems Conferences, vol. 1, no. 3, pp. 34–46 (2015)
111. Vaslin, R., Gogniat, G., Diguet, J., Wanderley, E., Tessier, R., Burleson, W.: A security approach for off-chip memory in embedded microprocessor systems. Microprocess. Microsyst. **33**(1), 37–45 (2009). https://doi.org/10.1016/j.micpro.2008.08.008
112. Standard, M., Fields, P., Alrimeih, H., Rakhmatov, D.: Fast and Flexible Hardware Support for ECC Over, vol. 22, no. 12, pp. 2661–2674 (2014)
113. Fons, M., Fons, F., Cantó, E.: Embedded security: new trends in personal recognition systems. IEEE Commun. Mag. **6**(1), 89–92 (2012)
114. Babar, S.D.: Security framework and jamming detection for internet of things: a dissertation submitted to the Department of Electronic System of Aalborg University in partial fulfillment of the requirements for the degree of doctor of philosophy. [Online]. http://vbn.aau.dk/files/208811916/Thesis_SECURITY_FRAMEWORK_AND_JAMMING_DETECTION_FOR_INTERNET_OF_THINGS.pdf (2015)
115. Liu, T., Guo, H., Parameswaran, S., Hu, S.X.: Crossmark. Integr. VLSI J. **56**(April 2016), 96–104 (2017). https://doi.org/10.1016/j.vlsi.2016.10.006

116. Nam, H., Lysecky, R.: Latency, power, and security optimization in distributed reconfigurable embedded systems. In: IEEE International Parallel and Distributed Processing Symposium Workshops, pp. 0–7 (2016). https://doi.org/10.1109/IPDPSW.2016.40
117. Chehab, M.I., Abdallah, A.E.: Architectures for identity management. In: 2009 International Conference for Internet Technology and Secured Transactions, {ICITST}, pp. 1–8 (2016). 978-1-4244-5647-5. https://doi.org/10.1109/ISCC.2018.8538636
118. Peters, M.E., Neumann, M., Zettlemoyer, L., Yih, W.T.: Dissecting contextual word embeddings: architecture and representation. In: Proceedings of the 2018 Conference on Empirical Methods in Natural Language Processing. EMNLP 2018, pp. 1499–1509 (2020). https://doi.org/10.18653/v1/d18-1179
119. Gelbart, O., Leontie, E., Narahari, B., Simha, R.: A compiler-hardware approach to software protection for embedded systems. Comput. Electr. Eng. **35**(2), 315–328 (2009). https://doi.org/10.1016/j.compeleceng.2008.06.010

Chapter 3
Embedded IoT Security Framework

This chapter presents the proposed security framework for IoT networks. The security vulnerabilities of the existing solutions led to the developing of a security framework for IoT networks. The framework was developed using a cost-effective design approach due to the constrained nature of IoT devices. Two components are used in developing the protocol, i.e., capability design (mainly a ticket, token, or key that provides authorization to access a device) and Advanced Encryption Standard (AES)-Galois Counter Mode (GCM) (a security protocol for constrained IoT devices). AES-GCM is an encryption process based on authentication and is suited for IoT. The novelty of this approach is that it improves IoT network functionality in terms of energy consumption, traffic delay, and throughput during the communication process. Since data communication in this framework is carried out wirelessly, reactive jamming attacks are introduced in the communication channel to ascertain the effectiveness of reactive jamming attacks in IoT networks. It also discusses reactive jamming modeling and attack formation in IoT networks. The reactive jamming model captures all the strategies followed by a malicious attacker. The key attributes of these models lie in their simplicity and effectiveness. The model is categorized into two, namely, activity and behavioural (sequential) models. This research presents both models using Unified Modelling Language (UML) to provide a detailed understanding of reactive jamming attacks. We use the game theoretical approach that provides detailed interaction among nodes for formulating reactive jamming attacks. Game theory is a branch of applied mathematics that provides a formal modeling approach to study strategic interactions and decision-making in competitive environments [1]. We formulate a reactive jamming attack game between two players: "a non-cooperative" and "zero-sum game" to analyze the interaction between the attacker and monitoring nodes in the network.

F. A. Alaba and A. Rocha, *Security Framework and Defense Mechanisms for IoT Reactive Jamming Attacks*, Studies in Systems, Decision and Control 548,
https://doi.org/10.1007/978-3-031-65929-4_3

3.1 System Model

In IoT, the possible communications are device-to-device, human-to-device, and human-to-human, which connect heterogeneous entities or networks. This section discusses the entire system model for reactive jamming attacks in IoT. The system model enables us to understand the sequence of activities that occur when reactive jamming attacks happen in an IoT network. The system model also helps to understand an actual view of the IoT networks and enables us to decide on the mitigation plans [2]. Figure 3.1 presents the reactive jamming hierarchy system model.

Figure 3.1 illustrates the hierarchical model for reactive jamming attacks in the IoT network's physical (PHY) layer. The model comprises four major components: reactive jammer, jamming detection/mitigation, system utility module, and PHY layer driver.

The reactive jammer components implement the reactive jamming strategies in the network. It depends on the system utility module and provides interfaces for the system utility module. The jamming detection/mitigation components provide and implement reactive jamming detection/mitigation strategies along the communication channel in the network. This feature also provides interfaces and depends on the system utility module. The system utility module bridges the jammer layer and the PHY layer. It offers a set of functions for reactive jammer and jamming mitigation classes to utilize for implementing their strategies. Finally, the PHY layer driver

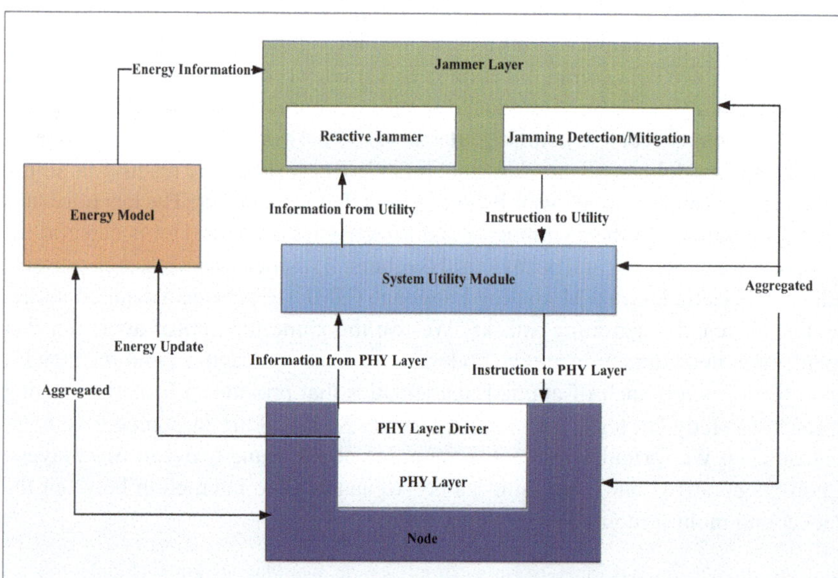

Fig. 3.1 Reactive jamming model

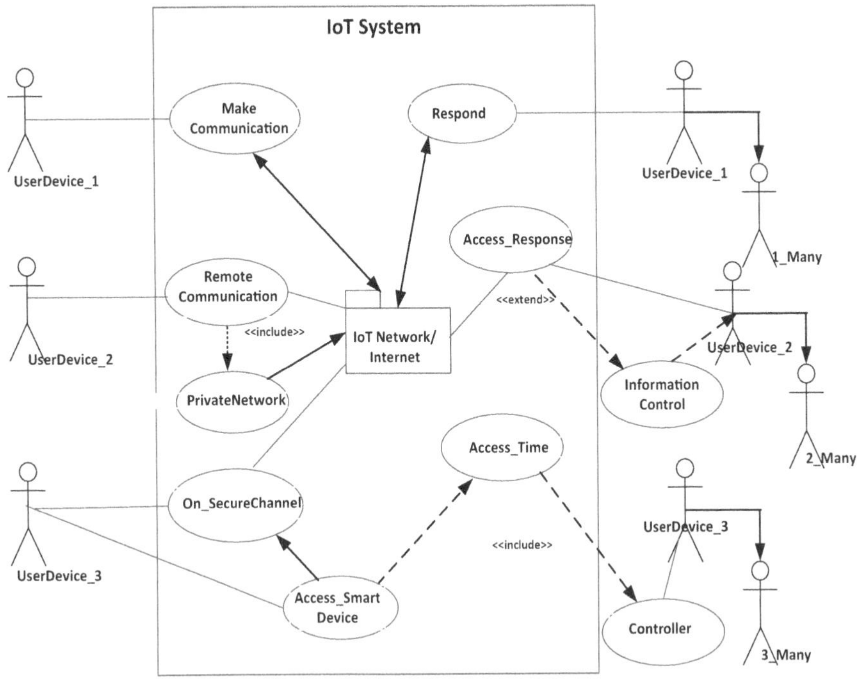

Fig. 3.2 IoT attack communication scenario using use case

modifies the PHY layer, and the provision of the PHY layer interfaces to the system utility module.

To better understand the reactive jamming hierarchical model in IoT in the presence of attacks, we illustrate a self-explanatory scenario between mobile devices and how communications occur using a "Use Case" approach, as shown in Fig. 3.2.

Figure 3.2 shows the IoT Attack Communication Scenario using the Use Case. UserDevices 1, 2, and 3 represent three different and most probable scenarios in the communication system [3]. A mobile device represents an entity, i.e., any device in the network that communicates with other entities of the same type or different type via the Internet or directly.

3.2 Basic Factors for Improving IoT Security Framework for Reactive Jamming

The essence of developing an IoT security architecture for reactive jamming attacks is ensuring that the security protocols and mechanisms are adequately embedded inside IoT devices from the development stage [4]. Embedding robust security requirements at the building stage will prevent devices from being exposed to threats and attacks like DoS (jamming) during device-to-device communication [5]. Therefore, it is essential to consider the security requirements during the design process of embedded IoT devices. The reactive jamming attacks security modeling is illustrated in Fig. 3.3.

Figure 3.3 shows the reactive jamming attack security modeling and the network vulnerability to attacks. All devices in IoT have low memory and limited computation resources, making them prone to resource enervation attacks. Attackers can send messages or requests to specific devices to consume their resources. This attack is more daunting in IoT because an attacker might be single in number and resource-constrained devices are large in numbers, leading to more energy consumption on communicating nodes and traffic delay. DoS (Jamming) attack is also possible due to man-in-the-middle attack. For this purpose, authentication and access control are the main security issues that have to be addressed. This chapter presents an integrated lightweight solution for authentication and access control with protocol evaluation for IoT.

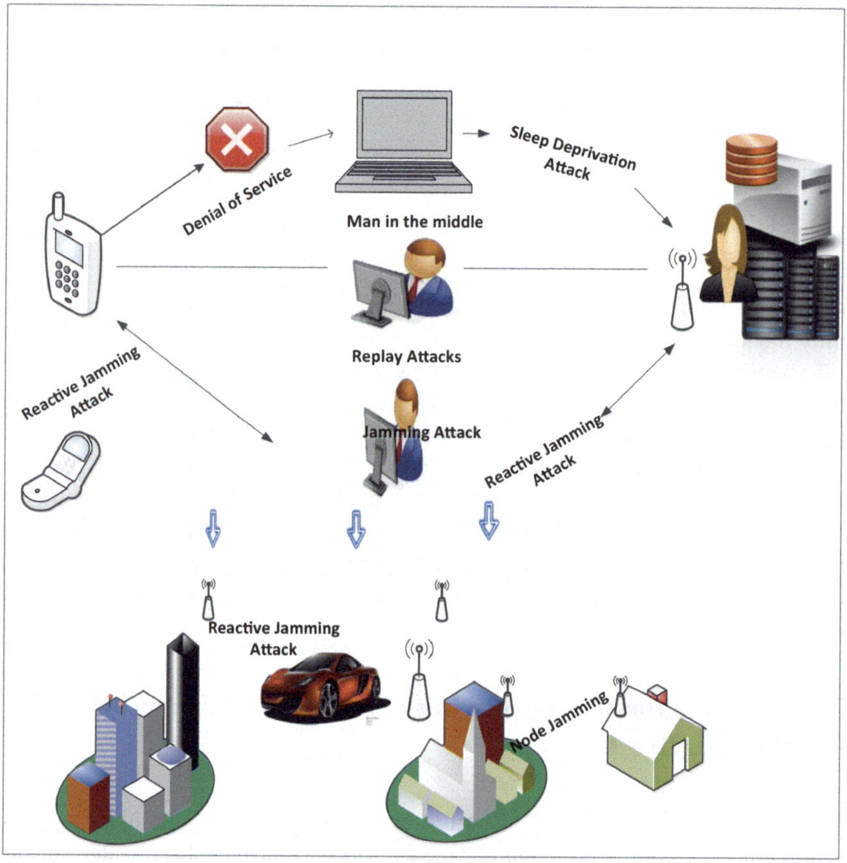

Fig. 3.3 Reactive jamming attacks security modelling

The basic factors that need to be considered to improve the security framework for embedded IoT devices are:

a. **Security Requirements**: Every security requirement should address a specific security need and knowledge of the vulnerabilities that could exist in an application. Generic guidance and knowledge is not enough. Specific applications have specific security requirements. Security requirements are Non-Functional Requirements (NFRs) that define the functional system security requirements, such as integrity, confidentiality and availability. Non-functional requirements describe what a system has to be. These statements support auditability and uptime. An example of a Non-functional security requirement statement is "Audit logs shall be verbose enough to support forensics."
b. **Security Objectives**: The objective of cryptography-based security is to protect information resources by making unauthorized acquisition of or tampering with the information more costly than the potential value that might be gained. Because the value of information usually decreases over time, good cryptography-based security protects information until its value is significantly less than the cost of illicit attempts to obtain or tamper with the information. Good cryptography, when properly implemented and used, makes attempts to violate security cost-prohibitive. Security objectives focus on system software development from the starting phase to prevent threats and vulnerabilities.
c. **Environmental Security Factor**: This helps to monitor and analyze threats, attacks and vulnerabilities posed on the system-operating environment. There are numerous IoT applications, but most systems require an environment that is resource inexpensive to simulate.

Figure 3.4 shows the basic requirements for an IoT security framework, which should be considered when developing a security framework for embedded IoT devices to build functional security policies to improve security for embedded devices.

3.3 Proposed Embedded Security Framework and Architecture

The proposed embedded security framework focused on the software platform of the IoT network. The software platform uses lightweight authentication protocols as the standardized protocols for securing resource-constrained devices. Lightweight protocols provide the necessary security support for IoT network layers. The architecture for embedded devices should be able to offer physical security to secret keys by securing the following components:

a. Secure ROM deals with secret keys.
b. Secure Boot Loader ensures that the genuine OS is used for booting the device.

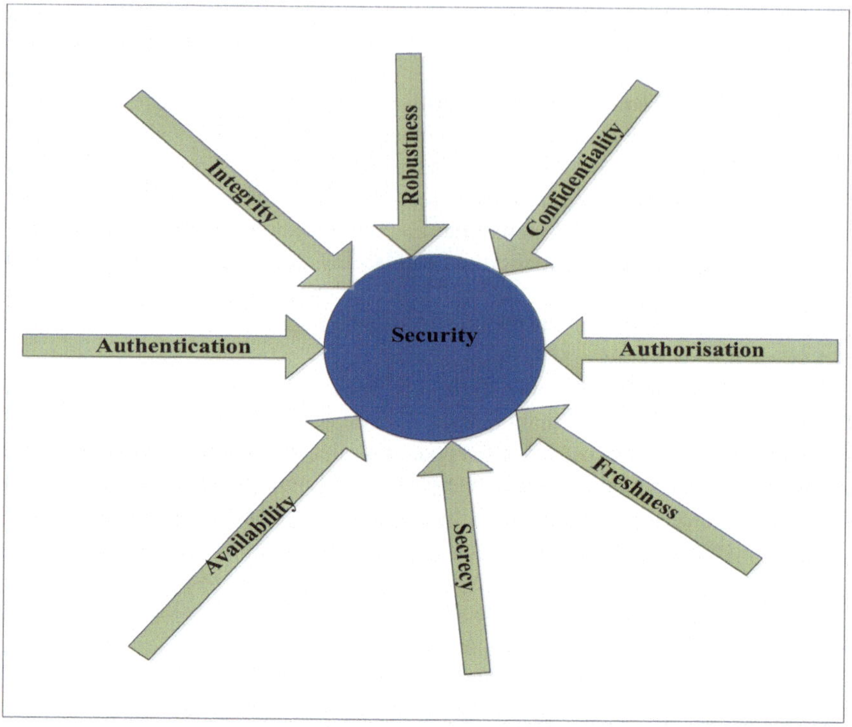

Fig. 3.4 Basic requirements for security framework for IoT

To develop inbuilt security for embedded IoT devices, the major components that need to be considered are secure ROM, secure memory management unit, secure runtime execution environment, and standard security protocols. All these will improve the design of a protected security architecture and framework for IoT.

The key components of the secure architecture are secure communication in the processor, secure software management, and secure hardware blocks. Failure to secure these components will make providing adequate security for IoT devices difficult.

Similarly, the main features of the security framework are;

a. Lightweight cryptography is an enhanced cryptographic algorithm for resource-constrained and low-power devices.
b. Secure OS that provides secure communication, secure booting, and secure contents.
c. Standard security protocols provide cryptographic computations and are lightweight in terms of communication.
d. Defense against different types of attacks.

e. The physical security checks the security level in the physical layer, and consider the hardware device's vulnerabilities.
f. A secure storage for securing stored data in either RAM or ROM.

Figure 3.5 shows the proposed framework developed using a cost-effective design approach. A cost-effective design approach uses hardware and software components to achieve security goals and build a cost-effective system. It is an approach that provides adequate security when implementing security for resource-constrained devices. This design approach enables system analysts to develop cost-effective systems and design system trade-offs.

This research focuses on the software platform only due to cost implications regarding getting IoT hardware devices. However, the design approach also considers the hardware platform. Researchers can explore The hardware platform in the future, which can lead to the unification of the IoT security framework when both platforms are combined. It is stated in Chap. 2 that IoT lacks a unified security framework and architecture, which makes it not acceptable to users due to its vulnerability to different attacks. The security framework of embedded IoT devices specifically depends on system specification, device size, and network specification. A detailed understanding of these parameters will help provide the required information for defining the framework's limitations.

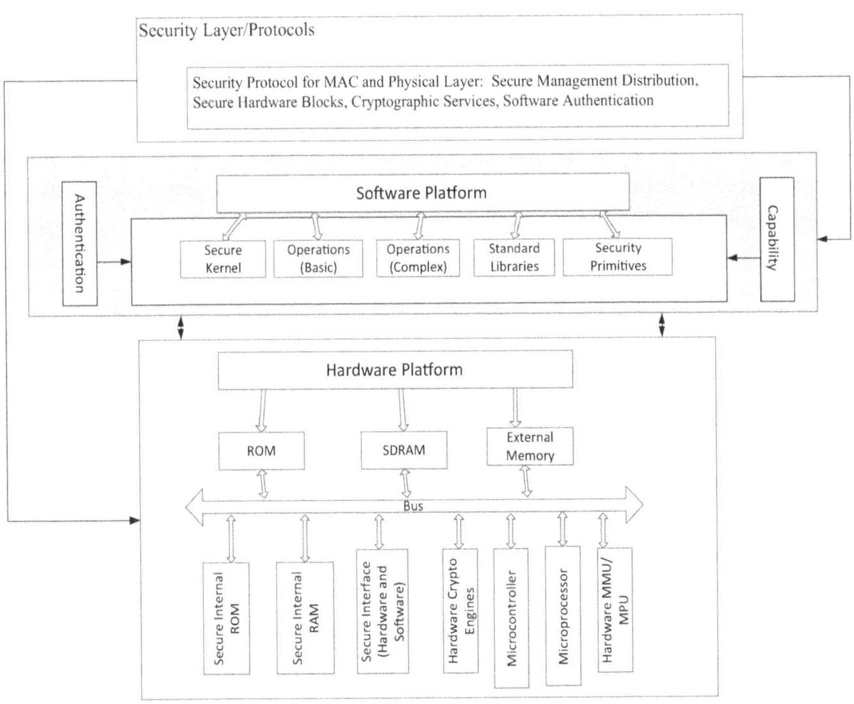

Fig. 3.5 Proposed security framework for embedded IoT

3.4 Proposed Authentication and Encryption Scheme for IoT Framework

Lightweight cryptographic encryption is the latest concept commonly used for securing embedded IoT devices. It helps provide authentication and data encryption that can easily be deployed as a security scheme for embedded devices. In this research, we use the Advanced Encryption Standard (AES)-Galois Counter Mode (GCM) technique for the authentication and encryption process in the IoT. The GCM is a security protocol for constrained IoT devices. It is a strong security model for message authentication and confidentiality against attacks introduced by [6]. The GCM uses block cipher mode for an operation with universal hashing in a finite field Ghash Function (GF) (2^{128}) to provide message authentication encryption. To achieve low cost, high speed, and minimum latency, GCM should be implemented in the IoT software platform during the development stage, which provides a good device performance. It reduces the energy consumption when nodes are authenticating, reduces the traffic delay during communication, and, thus, improves the network throughput [7]. It uses tools that are well-understood and supported by a theoretical basis. Its security depends on the security of the block cipher. It is a normal assumption in cryptographic designs, and it also appears to be valid for AES.

The AES-GCM is the most recent authentication encryption algorithm suitable for resource-constrained devices. It provides message confidentiality and authentication for embedded devices [7]. The AES-GCM authenticated encryption operation has four inputs, each input in the form of a string:

a. A secret-key K, whose length is suitable for the basic cipher block.
b. Initial Vector (IV) can have any number of bits between 1 and 2^{64}. The major purpose of IV is to be a nonce. For a constant value of key, every IV value needs to be unique.
c. A plaintext P can have any number of bits between 0 and 2^{39} to 2^{56}.
d. Additional Authenticated Data (ADD), which is denoted as A. "A" implies that data is authenticated but not encrypted. It can be any random number of bits between 0 to 2^{64}.

To provide strong security encryption and confidentiality, AES-GCM utilized the AES block cipher in counter-mode (i.e., a mode of operation that repeatedly applies a cipher's single-block operation securely to transform a large amount of data that is larger than a block) to provide high-level security that is well-suitable for hardware implementation [8]. In this manner, the best security solution for embedded IoT devices is the AES-GCM technique.

In addition, AES-GCM meets the necessary security requirements for resource-constrained devices. It improves the computational speed and storage capacity when implemented using a hardware and software co-design approach [9]. The proposed protocol makes use of capacity-based addressing in [9–11] in addition to AES-GCM to provide access control to the devices.

3.5 GCM Notation

In this research, we adopt the recommended "block cipher modes of operation" notations by [7]. GCM has major functions, block cipher encryption, and multiplication, used over the field GF (2128) for robust authentication.

The block cipher encryption is denoted as $E(K, A')$, where A is the encryption value and K is the key. The multiplication and addition of two elements (i.e., A' and B') are denoted as $A' \cdot B'$ and $A' \oplus B'$ respectively, given that $A', B' \in GF\ (2^{128})$. The concatenation of two-bit strings X and Y is denoted as $X \| Y$. The function len() returns a 64-bit string containing the non-negative integer describing the number of bits in its argument, with the least significant bit on the right. The expression 0^l denotes a string of l zero bits. The function $\mathrm{MSB}_t(S)$ returns the bit string containing only the most significant (leftmost) t bits of S, and the symbol { } denotes the bit string with zero length.

3.5.1 *Encryption*

Plaintext consists of a sequence of bit strings between 0 to 128. Given that, x and y denote a pair of unique positive integers, such that the total number of bits in the plaintext is $(x-1)128 + y$, where $1 \le y \le 128$. The data block is denoted as J^*_n with the following bit sequence $J_1, J_2 \ldots J_{n-1}, J^*_n$. Similarly, $C^*_n \in y$ represents the ciphertext with bit strings sequence of $C_1, C_2, \ldots, C_{n-1}, C^*_n$. In addition, ADD "A" has the following string sequence of $A_1, A_2, \ldots, A_{m-1}, A^*_m$. In this case, the last string, A^*_m, is a partial block of length r. Both m and r are unique positive integers such that the total number of bits in A is $(m-1)128 + r$ and $1 \le r \le 128$. The GHASH function is defined as $GHASH(H, A, C) = A'_{m+x+1}$.

Therefore, we define the authentication encryption operational process using Eq. (3.1) and Fig. 3.6.

$$H = E\left(K, 0^{128}\right)$$

$$B'_0 = \begin{cases} IV \| 0^{31} 1, & if\ len(IV) < 96 \\ GHASH\ (H, \{A\}, IV) & Otherwise. \end{cases}$$

$$B'_i = incr(B_{i-1}) \quad for\ i = 1, \ldots, n \tag{3.1}$$

$$C_i = P_i \oplus \mathrm{E}\left(\mathrm{K}, B'_i\right) \ for \ \ i = 1, \ldots, n-1$$

$$C^*_n = P^*_n \oplus MSB_y(\mathrm{E}\left(\mathrm{K}, B'_n\right))$$

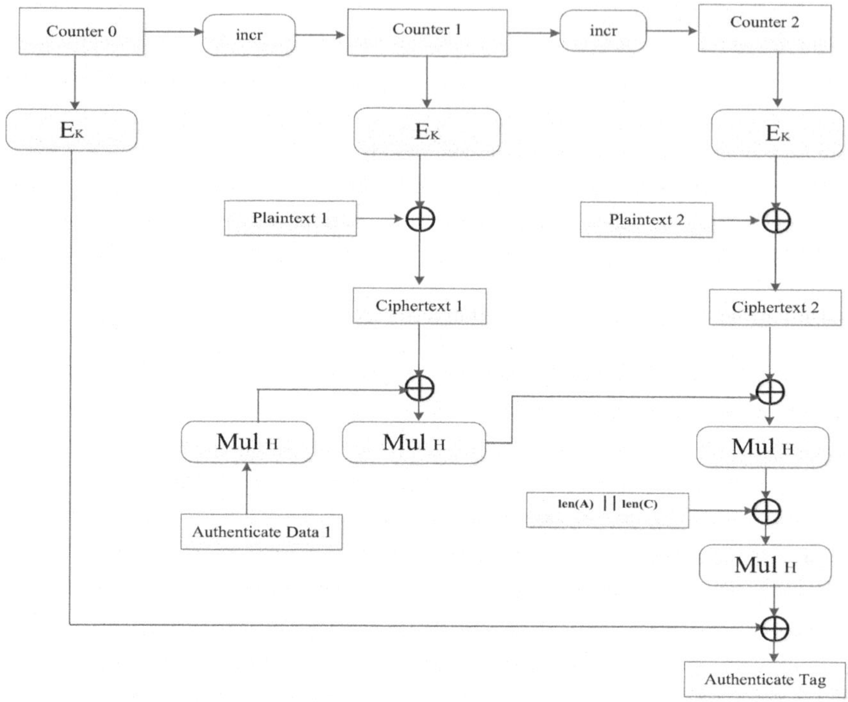

Fig. 3.6 Data encryption operational process

$$T = MSB_t(GHASH\left(H, A', C\right) \oplus \mathrm{E}(\mathrm{K}, B_0'))$$

Figure 3.6 shows the encryption operational process of GCM.

An authenticated data encryption operational process for a situation when only a "single block of additional authenticated data" (i.e., Authenticate Data 1) and plaintext of two blocks is provided in Fig. 3.3. E_K denotes the block cipher encryption using the key K, Mul_H denotes multiplication in GF (2^{128}) by the hash key H, and incr denotes the counter increment function.

3.5.2 *Decryption*

The authenticated data decryption operational process is related to the operational process of data encryption, but with the order of the hash and encrypt steps reversed. Thus, we define the authentication decryption operational process using Eq. (3.2) and Fig. 3.7.

$$H = E\left(K, 0^{128}\right)$$

$$B'_0 = \begin{cases} IV \| 0^{31}1, & \text{if } len(IV) < 96 \\ GHASH\,(H, \{A\}, IV) & Otherwise. \end{cases}$$

$$B'_i = incr(B_{i-1}) \quad fori = 1, \ldots, n \tag{3.2}$$

$$P_i = C_i \oplus \mathrm{E}(\mathrm{K}, B'_i) \quad fori = 1, \ldots, n$$

$$P^*_n = C^*_n \oplus MSB_y(\mathrm{E}(\mathrm{K}, B'_n))$$

$$T' = MSB_t(GHASH\,(H, A', C) \oplus \mathrm{E}(\mathrm{K}, B'_0))$$

The tag T' that is computed by the decryption operation is compared to the tag T associated with the ciphertext C. If the two tags match (in both length and value), then the ciphertext is returned. Otherwise, the special symbol **FAIL** is returned. Figure 3.4 shows the operational process of data decryption.

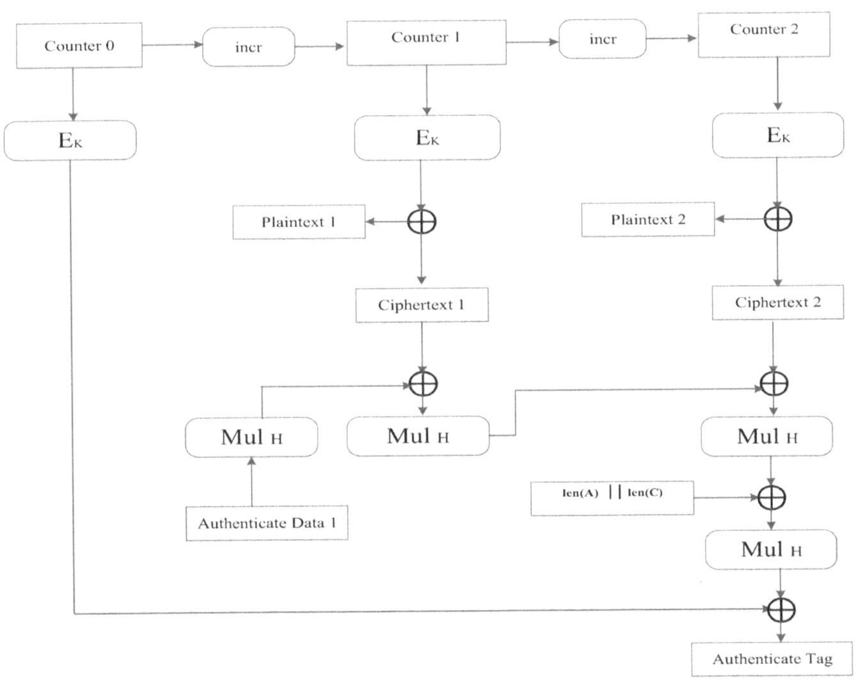

Fig. 3.7 Data decryption operational process

Table 3.1 NGF (2^{128}) multiplication

Algorithm 3.1: GF (2^{128}) Multiplication

Required

$D \in GF(2^{128})$; $D \leftarrow 0$, $R \leftarrow K$

Input: K, L

Output: D, R

While $0 \leq i \leq 127$ **Do**

  If $B_i = 1$ **Then**

  $D \leftarrow D \oplus R$

Elseif

  if $R_{127} = 0$ **Then**

  $R \leftarrow$ rightshift(R)

  Else

  $R \leftarrow$ rightshift(R) $\oplus L$

  End if

End

Note The function rightshift() moves the bits of its argument one bit to the right

3.5.3 GF (2128) Multiplication

The multiplication operation is defined as an operation on bit vectors to simplify the specification. Each element is a vector of 128 bits. The ith bit of an element K is denoted as K_i. The leftmost bit is K_0, and the rightmost bit is K_{127}. The multiplication operation uses the special element $R = 11{,}100{,}001 \| 0^{120}$, defined in Algorithm 3.1, presented in Table 3.1.

AES-GCM pre-processed packets before transmission to prevent jamming, but the packets remain unencrypted. Thus, the jammer cannot perform packet classification until all pseudo-messages corresponding to the original packet have been received, and the inverse transformation has been applied. AES-GCM is a publicly known and completely invertible pre-processing step to the plaintext before passing it to an ordinary block encryption algorithm.

3.5.4 The Proposed Protocol

This research proposes a "capability-based authentication and access control protocol" for embedded IoT devices. AES-GCM is then used to encrypt the data to provide strong encryption and authentication for the IoT devices. It is a protocol that provides mutual authentication among devices. It additionally addresses security issues related to jamming attacks and capability-based access control. The efficiency of the proposed protocol is tested by introducing a jamming attack in the communication channel. A jamming attack is the most dangerous type in the wireless communication medium. Before WSNs were generally and largely accepted, jamming attack was the first security issue cured in the network. Therefore, reactive jamming attack (i.e., being the most dangerous among other types of jamming attacks) for IoT network is also introduced in the communication channel. A detailed behavioral and activity model of reactive jamming attacks is presented in Chap. 4.

Capability is a ticket, key, or token that provides authorization to access a device. Its implementation is in the form of a data structure composed of an identifier, a random number, and access rights. On IoT devices, a single name is used as an identifier. Sensors are used to equip IoT devices. The operations to be implemented on the devices are outlined by access rights.

Figure 3.8 shows a standard capability structure comprising the following attributes: Device, Rights, and Random. All these attributes are represented as a single ticket for capacity. The device identifier is the name or identity of a device. Access rights are sets of access information that describe device capability. Thus, the capability is denoted as;

$$Capability = (ID_n, R_n, RNG) \tag{3.3}$$

From Eq. (3.3), ID_n represents sets of device identifier, R_n represents set of access rights, and RNG represents the random number generator.

Access rights (R_n) can be read, write or null object/data. If access right (R_n,) then permission is granted to access that particular object. It is defined in Eq. 3.3:

$$R_n \in \{Read, Write, NULL\} \tag{3.4}$$

A random number generator (RNG) generates random numbers using one-way hash function in order to avoid devices/data forgery. A one-way function is used with a random number to detect interference for an access request and a capability that

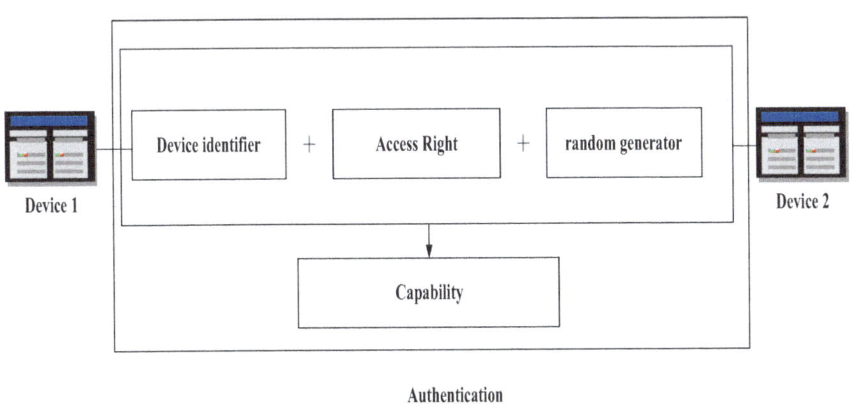

Fig. 3.8 Capability structure for IoT devices

Table 3.2 Notations and descriptions

Notation	Description
ID_1	Identifier of the first device 1
ID_2	Identifier of the second device 2
R_1	Access rights of the first device 1 "D_1"
R_2	Access rights of the second device 2 "D_2"
rn	Random number generated by D_1
K	Secret key
H()	One-way hash function
E()	Encryption using AES-GCM
RNG	Random number generator

consists of device identifier. Thus, access is granted if the capability is valid. RNG can be computed using Eq. (3.5).

$$RNG = f(ID_n, R_n) \quad (3.5)$$

The algorithm for the one-way hash function is made public and independent from the secrets keys, because many challenges unsuitable for resource-constrained devices are encountered during key distribution. One-way hash function is used based on the following advantages:

a. It is not possible to reverse a hash function.
b. When a set or pair of input and their equivalent output is given, the first input and its equivalent output is known but it is impossible to know the second input and its equivalent output.

Table 3.2 shows the notations and descriptions of all schemes used in the proposed protocol.

3.6 Operational Process for the Proposed Protocol

This section provides the operational process for the proposed protocol. Two components are used in developing the protocol: capability design and AES-GCM Scheme. Figure 3.9 illustrates the operational process of the protocol in detail. From Device 1, a capability is created, which is a function that contains both the device name and the access rights. The capability is encrypted using the AES-GCM algorithm to avoid forgery and hashed with a random number. A ciphertext is created during this process and sent to device 2 in encrypted form. The capability of device 1 is received and decrypted by device 2 using a symmetric key. A one-way hash function is used for verification purposes to check if the ciphertext has been tampered with using the following procedures:

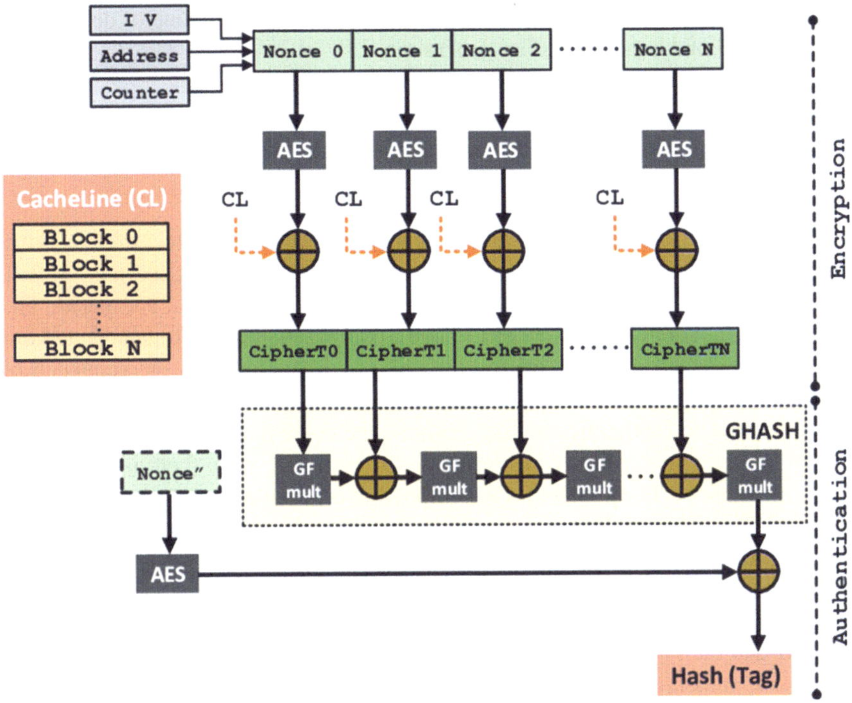

Fig. 3.9 Proposed protocol

a. Suppose the hash value generated is not equal to the hash value received. In that case, it is obvious that the communication channel has been tampered with, and this interrupts the authentication process.
b. Suppose the hash value generated equals the value received after the decryption process. In that case, it shows no interference, and authentication will now take place between device 1 and device 2.

The proposed protocol allows device 2 to authenticate itself with device 1 mutually. By so doing, device 2 performs its capability through the same process as device 1. In order to avoid replay attacks, the same random number is used by both devices. After the response have been received by device 2, the ciphertext is now decrypted to check for integrity. The random number between both devices is compared to confirm whether the message came from the right source. When the decryption and comparison is successful, device 2 is now authenticated to device 1. Both devices can now communicate with each other freely over a secure channel.

Note that the access rights of both devices securely communicate with each other in order to achieve a secure access control.

The proposed protocol integrates the AES-GCM technique to provide authentication and encryption for resource-constrained devices with minimum latency, low operational overhead, and efficient, low-cost implementation, which is unavailable in the previous protocols. It also uses the capability as a second line of defense for access control and random number generators.

3.6.1 Using AES-GCM

This section discusses the importance of using the AES-GCM protocol for securing IoT devices. It illustrates how packets/data are encrypted and decrypted in IoT networks. During encryption, the data fields are authenticated and encrypted. The encrypted data is conveyed with a header and a number sequence. The header authentication is done by incorporating AAD, while IV is included in the sequence of numbers. In addition, the Integrity Check Value (ICV) field is included alongside the encrypted data and authentication tag. It should be noted that, during the encryption process, there is no need to reduce/mitigate the plaintext, since there is no specific length for an input. The plaintext is the output that is provided during the authentication decryption process. However, in a situation when the authentication check fails, the decryption process instead of plaintext will return FAIL. The decapsulation will be terminated, and the plaintext will be discarded.

After all these processes/operations, the header and sequence number can be checked, and their values can be trusted. The AES-GCM encryption and decryption process for authenticating IoT devices is illustrated in Figs. 3.10 and 3.11, respectively, which uses cryptographic encapsulation and IEEE 802 Media Access Control Security draft standard [12, 13].

Figures 3.10 and 3.11 elucidates how fields of the security encapsulation map onto the inputs and outputs of the authenticated encryption and decryption mode. In some situations, having the same AES-GCM key used for encryption by more than one device may be desirable. In this case, coordination is needed to ensure the

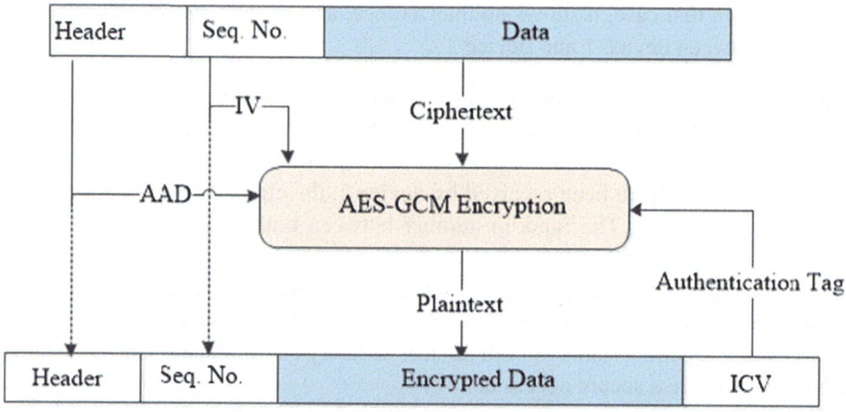

Fig. 3.10 AES-GCM encryption and authentication process for IoT devices

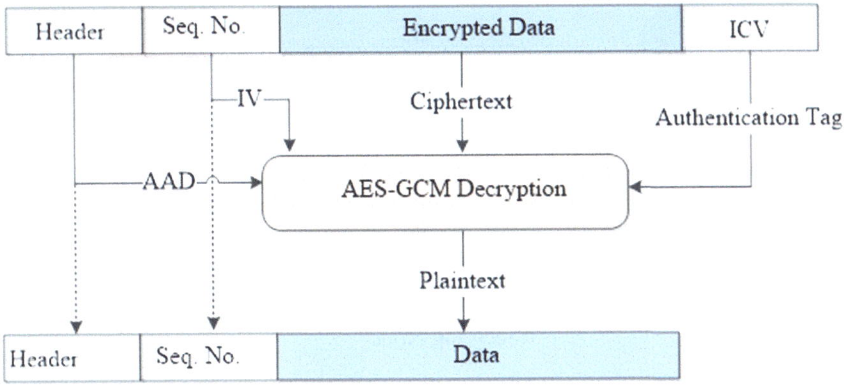

Fig. 3.11 AES-GCM decryption and authentication process for IoT devices

uniqueness of the IV values. A simple way in which this requirement can be met is to include a device-specific value in the IV, such as a network address. In subsequent sections, the reactive jamming attack modeling and formation in IoT networks will be provided.

3.7 Reactive Jamming Attack Modelling

This section provides reactive jamming attack activities and behavioral models in IoT networks using activity and sequential modeling techniques. The activity and sequential modeling techniques provide a detailed understanding of reactive jamming attack behavior, i.e., it shows the level of data flows, objects involved, and how messages are exchanged among various entities in the network during the interaction. In addition, the activity modeling provides a high level of understanding of the device functionalities and describes the data flow among users and devices. Sequential modeling provides dynamic interaction between objects at a low level and demonstrates how messages are exchanged among objects during data communication for reactive jamming. The system model of a sender node, receiver node, and reactive jammer is illustrated in Fig. 3.12.

Reactive jamming is only feasible when the system's geometry is such that the jammer's transmitted signal reaches the target receiver before it hops to a new channel or stops transmitting. Reactive jamming is only possible when the jammer is physically located near or between the target transmitter and receiver. If λit represents the fraction of each node duration that must remain not jammed for communications to succeed, then from Eq. (3.6), we have the following inequality limiting the distances S_2 and S_3;

$$S_2 + S_3 \le (\lambda R_s - R_j)q + S_1 \tag{3.6}$$

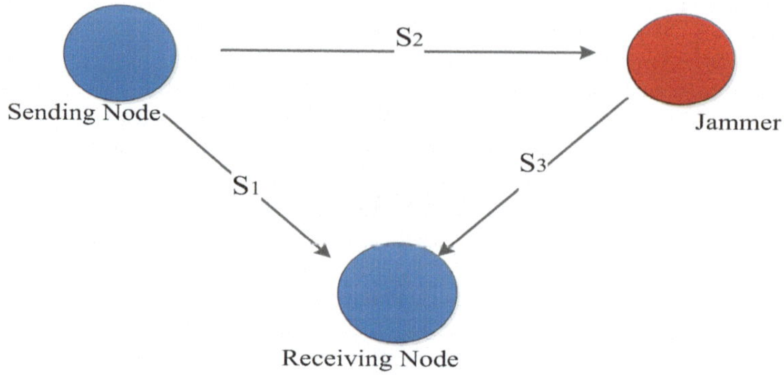

Fig. 3.12 System model for sender node, receiver node, and reactive jammer

where R_s is the node duration, R_j is the jammer's processing time, q is the speed of light, and S_1, S_2, and S_3 are the distances indicated in Fig. 3.12.

The reactive jammer continuously monitors the communication medium and immediately transmits a radio signal upon sensing a packet transmission to cause a collision at the receiver. The jammer can transmit at just enough power to corrupt a single bit to generate a received packet to fail the Cyclic Redundancy Codes (CRC) check. Normally, a reactive jamming attack has the following criteria: high energy efficiency (i.e., consumes low power), low probability of detection (preferably close to 0), achieves high levels of DoS (i.e., disrupts communications to the desired (or maximum possible) extent and be resistant to PHY layer anti-jamming techniques (i.e., does not allow signal processing techniques to overcome the attack). Often, the criteria of interest are dependent on the jamming scenario. In other words, the jamming scenario dictates the most suitable criteria for use. For instance, when malicious nodes have limited energy resources, energy efficiency will be their prime goal. Based on these criteria, different jammers tend to be effective in all cases as long as the criteria above are possible. For example, to maintain a low probability of detection, the jammer can adopt techniques consistent with MAC layer behaviors.

3.8 Activity Model for Reactive Jamming Attacks

Activity models describe the functional understanding of any system through the demonstration of all logical processes involved during communication. In an activity model, each logical process represents a sequence of tasks along with decisions that manage the performance of each task. The activity modeling uses UML to illustrate the functional view of tasks or processes and aims to describe each task's behaviors [13]. The flow in an activity diagram is motivated by the accomplishment of an action. The activity diagram is a valuable instrument that provides a detailed understanding of the basic flow of security attacks. The accomplishment of any task/

process determines the action and data flow in an activity diagram. In addition, an activity diagram is a very good tool that helps provide a detailed understanding of the basic flow security attacks in systems. The activity modeling of reactive jamming attacks in IoT networks is provided in the following sub-section.

3.8.1 Reactive Jamming Activity Modelling

Reactive jamming is the most challenging attack to detect in an IoT network. It is more dangerous than constant, deceptive, and random jamming attacks in terms of performance [14, 15]. In reactive jamming, the attacker turns a normal node into a jammer by transferring the jamming features to the victim, which jams another node's transmission [16]. A reactive jammer transmits signals only when it detects that the normal node sends packets to other nodes. It keeps quiet. This makes detecting reactive jamming difficult [17]. The model tries to break down the communication between nodes. While they can achieve a high degree of DoS, they exhibit (in general) low energy efficiency and a high probability of detection. The activity model of this type of jammer is shown in Fig. 3.13.

Figure 3.13 demonstrates the activity model for reactive jamming and the various steps carried out by the reactive jammer when a jamming attack is to be executed. In

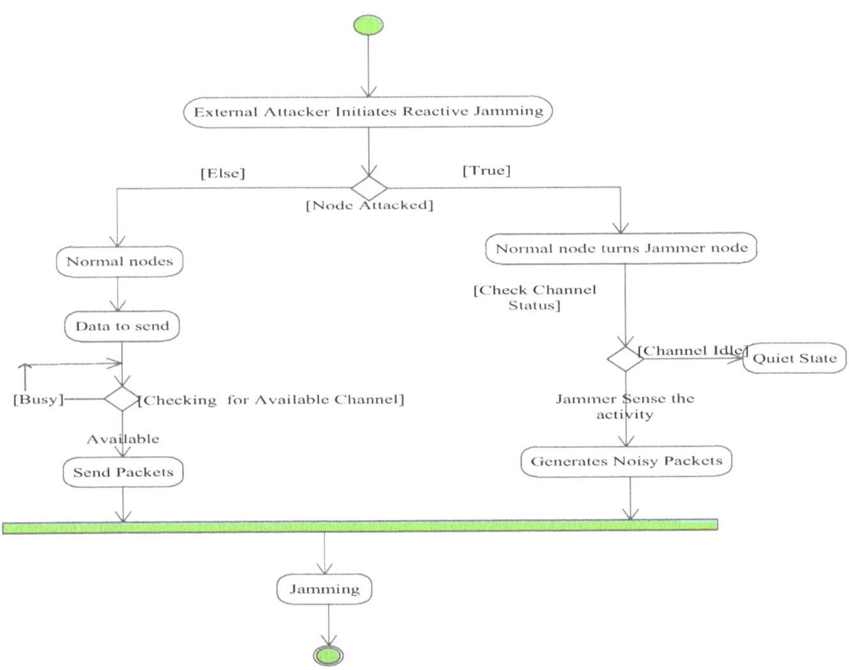

Fig. 3.13 Activity model for reactive jammer

reactive jamming, the jammer (attacker) sends a jamming signal across the communication channel when the normal signal is also sent, which turns a normal node into a jammer. The reactive jamming transmits signals only when it detects that the normal node is sending packets to other nodes; it keeps quiet.

The activity sequence of reactive jamming is as follows:

a. Initialization of reactive jammer: The key feature of reactive jamming is that it is initiated once the communication channel is busy when different nodes are trying to send data. The initialization is when the jammer node can sense and detect the legitimate sending and target nodes. Once the attack initialization is successful, the communicating nodes (victims) will possess the features of a reactive jammer. Otherwise, the normal node performs its normal tasks.
b. Detection of available communication channel: At this time, the sender nodes check and identify if there is any available channel. If a channel is available, then the sender sends packets through it.
c. Checking the channel's status: The jamming node observes the status of the communication channel to determine if it is busy or idle. If the communication channel is busy, the jammer node is activated and generates noisy packets continuously, leading to jamming. If the channel is idle, the jammer node sleeps. It reduces network performance because of its effect on the communication channel.

3.9 Behavioural Model for Reactive Jamming Attack

To quantify the extent to which the reactive jammer affects packet transmissions between nodes in an IoT network, we need to understand its behaviour. This section discusses the behavioral model of various kinds of jamming attacks in IoT network using sequential modeling in UML. The UML is introduced to provide a detailed description of the interactions between nodes in a sequential order. This research uses the sequence diagram to demonstrate how reactive jamming occurs among objects in IoT networks sequentially. The sequence diagram comprises vertical parallel lines with different objects and procedures, horizontal arrows, and messages exchanged between objects. The various nodes in the network and attacker are considered as entities and the interaction between them is called processing. A node is considered to be in a normal state when there is a successful exchange of Send Data and Clear To Send (CTS) during data transmission. The adversary deploys a malicious node physically between two or more legitimate IoT nodes along the channel, which in turn takes control of all data flow between the nodes as well as their operation. In addition, any attack on nodes is initiated by an external attacker, and the initiated nodes are now converted to jamming nodes. The following sub-section provides the behavioural modeling of reactive jamming attacks in IoT network.

3.9.1 Behavioral Model for Reactive Jamming Attack

In this section, we present jamming techniques involving the jamming of IEEE 802.11 control packets to corrupt communications with the minimum possible energy consumption. These techniques require a good knowledge and understanding of the IEEE 802.11 standard. The IoT devices are connected over an IEEE 802.11ah (low-power WiFi) protocol, which is known to be a suitable solution for resource-constraint IoT devices [18]. The reactive jammer targets the reception of a message and is harder to detect because it transmits only when it senses that the communication is active, by producing noise packets and causing huge amount of collision in the network as a result of its behaviour, and remains silent when the channel is idle. The behavioral model help us to understand the real operational behavior of reactive jamming attacks and provides a detail understanding of the objects involved during jamming and the interaction between them in IoT networks. The behavioural model of reactive jamming attack is presented in Fig. 3.14.

Figure 3.14 illustrates the reactive jamming attack, the sequence of interactions and actions between objects and nodes in the network. It is considered to be the most intelligent type of jamming attack that observes the network status before reacting. The CTS senses the channel for a malicious node and waits for a send data. After the transmission, it waits for a time interval and sends a short jamming pulse that results in corrupting the CTS packet. This strategy will result in zero throughput since no data will be successfully transmitted.

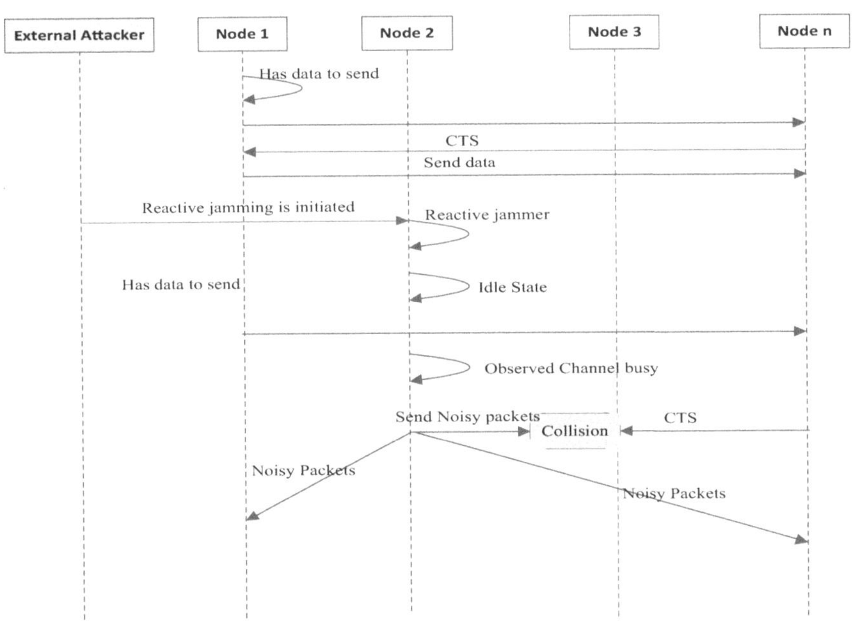

Fig. 3.14 Sequential model for reactive jamming attack

The interaction sequence and activities in the network are as follows;

a. Node 1 has data to send and checks if there is an available communication channel. If available, it sends the data to the destination node.
b. Reactive jamming is initiated on node 2 by the attacker, and the node is now a jammer node, which begins to analyze the network status (checking when the channel is busy and idle).
c. If the network is not busy, the reactive jammer becomes idle.
d. When node 1 (normal node) sends the data packet to node ***n***, the reactive jamming node observes the activity between the two nodes. It then sends noise packets to interrupt and cause a collision with CTS in the network. The attacker always initiates an attack when it observes any activity in the communication channel.

3.10 Reactive Jamming Model Formulation

The reactive jamming attack and detection model are formulated using game theory. Game theory offers models to capture the interaction between players, in this case, nodes, by modeling the players as components of social networks, where players can act in ways that maximize their utility, which does not always lead to favorable outcomes for the game. While game theory still lets players choose the best available action, it provides a situation where other player's utilities are also maximized. The detection model is formulated as a game between two players: "*a non-cooperative*" and "*zero-sum game*". A non-cooperative is used for mitigating security issues in IoT networks. The zero-sum game is a non-cooperative game between two players. One player is considered a "maximizer" and the other a "minimizer". The maximization strives to maximize its gain, while the minimizer's target minimizes its losses [19]. This approach is chosen because, in an IoT network, every node (player) tends to maximize their resources (gains) to be fully active during packet transmission. These players are monitoring nodes responsible for detecting the reactive jamming attack. The players are set as:

$$S = \{S1, S2\} \tag{3.7}$$

where $S1$ is the monitoring node, and $S2$ is the jammer.

The tradeoffs for the model formulation are *constant monitoring (Mc)* and *periodic monitoring (Mp)*. The monitoring nodes observe the transmission channel's activities continuously or at a predefined time interval. However, the communicating nodes select monitoring nodes based on attack strategy. A reactive jamming attack strategy is adopted in this model, and it is denoted as "ReJ". The monitoring node uses two kinds of strategies *(Mc, Mp)* for monitoring the channel.

Now the strategy is denoted as:

$$X = X1 * X2 \tag{3.8}$$

where $X1 = \{Mc, Mp\}$ and $X2 = \{ReJ\}$. Meanwhile, $X1$ and $X2$ represent player 1 and player 2, respectively.

A player is the strategy the jammer node displays in the channel to achieve its target. A utility function shows the effectiveness of the monitoring node concerning attacks that are successfully and unsuccessfully detected during transmission. In the case of the monitoring nodes, the detection rate and false positive rate are two utility functions considered. A jammer utility prevents successful packet transmission and reduces network throughput by introducing DoS attacks.

The utility function *(U)* is given as:

$$\{U\} = \{U1, U2\} \tag{3.9}$$

where $U1$ and $U2$ are detection rate and attack gain (i.e., the attack is successfully detected or launched), respectively.

From Eqs. (3.7), (3.8), and (3.9), a reactive jamming strategy model for an IoT network is formulated for both Mc and Mp strategies.

$$Mc = A_d\left(G_a - P_{rej}\right), \left(G_d - P_c\right) \tag{3.10}$$

$$Mp = A_d\left(tG_a - P_{rej}\right), t\left(A_d G_d - P_p\right) \tag{3.11}$$

where A_d denotes the attack duration, t represents Mp time, G_d denotes attack detection gain, G_a represents the gain for an attack that is launched successfully, P_{rej} is the payoff for reactive jamming, P_c and P_p are both payoffs when Mc and Mp are used to detect the attack. Gain is when an attack is successfully detected or launched. Payoff is the cost for detecting or launching an attack.

Equations 3.10 and 3.11 show the reactive attack strategy for both Mc and Mp. The reactive jammer launches an attack only when it detects any activity in the communication channel. An attack duration A_d is defined as the period when a reactive jammer launches an attack on the channel. In the case of *Mc,* paying cost P_{rej} in the period of A_d will enable the reactive jammer to accomplish its gain G_a. Otherwise, the jammer gain G_d by paying the cost of constant monitoring P_c. In *Mp, the* reactive jammer can achieve gain G_a at time t by paying P_{rej} in each attack duration $A_{d,}$ otherwise the jammer gain G_d at each A_d by paying P_p after t.

3.11 Conclusion

In this chapter, a challenge-response type of protocol that improves computation overhead of IoT devices is proposed. To achieve a sound security solution for embedded IoT devices involve the introduction of security mechanisms on these devices right from the development stage. It is essential to consider the security requirements during the design process of embedded IoT devices in order to detect the vulnerabilities of the devices right from the development stage. When the sources of the vulnerabilities is known, then a defense mechanisms should be embedded inside the devices during the design stage. Therefore, AES-GCM protocol and capabilities was introduced for embedded IoT devices. The AES-GCM has been proven to be concrete security mechanism for resource-constrained devices. The novelty of this approach is that it provides efficient authentication and encryption standard for embedded IoT devices implemented at low cost. It reduces that level of energy consumption to IoT devices by speedy authentication process, thus, reduces the delay along the channel and improve overall network throughput. Also uses capability as a second line of defense for access control to provide right values for authenticated. The utilization of hash function ***h()***, random number $\boldsymbol{R_n}$ and secret value $\boldsymbol{K}$ provide efficient security in both static and dynamic IoT network. This chapter also provides reactive jamming attack formation using game theory approach. Detail understanding of jamming attack activity, behavioural models and attack formation will serve as a useful tool and a good strategy for developing an efficient defense mechanism against reactive jamming attacks in IoT networks.

References

1. Ai, Z.Y., Zhou, Y.T., Song, F.: A smart collaborative routing protocol for reliable data diffusion in IoT scenarios. Sens. (Switzerland) **18**(6), 1–21 (2018). https://doi.org/10.3390/s18061926
2. Villa, V., Bruno, G., Aliev, K., Piantanida, P., Corneli, A., Antonelli, D.: Machine learning framework for the sustainable maintenance of building facilities. Sustain. **14**(2), 1–17 (2022). https://doi.org/10.3390/su14020681
3. Deogirikar, J., Vidhate, A.: Security attacks in IoT: a survey. In: 2017 International Conference on I-SMAC (IoT in Social, Mobile, Analytics and Cloud)(I-SMAC), pp. 32–37 (2017). https://doi.org/10.1109/I-SMAC.2017.8058363
4. Saad, M., Bin Ahmad M., Asif M., Khan, M.K., Mahmood, T., Mahmood, M.T.: Blockchain-enabled VANET for smart solid waste management. IEEE Access. **11**(November 2022), 5679–5700 (2023). https://doi.org/10.1109/ACCESS.2023.3235017
5. Seungjin, L., Abdullah, A., Jhanjhi, N.Z.: A review on honeypot-based botnet detection models for smart factory. Int. J. Adv. Comput. Sci. Appl. **11**(6), 418–435 (2020). https://doi.org/10.14569/IJACSA.2020.0110654
6. Bellare, M., Kilian, J., Rogaway, P.: The security of the cipher block chaining message authentication code. **399**, 362–399 (2000)
7. Hori, Y., Satoh, A., Sakane, H., Toda, K.: Bitstream encryption and authentication with Aes-Gcm in dynamically reconfigurable systems national institute of advanced industrial science and technology (AIST). Adv. Inf. Comput. Secur. Proc. **53**(12), 261–278 (2012)
8. Hinsenkamp, L., Zhou, G., Michalik, H.: Efficient and high-throughput implementations of AES-GCM on FPGAs. IEEE Des. Test Comput. **12**(5) (2013)
9. Anggorojati, B., Mahalle, P.N., Prasad, N.R., Prasad, R.: Capability-based access control delegation model on the federated IoT network. IEEE Trans. Mob. Comput. **5**(3), 604–608 (2014)
10. Jayasinghe, D., Ragel, R., Ambrose, J.A., Ignjatovic, A., Parameswaran, S.: Advanced modes in AES : are they safe from power analysis based side channel attacks. IEEE Internet Things J. **5**(14) (2014)

11. Sankaran, S.: Lightweight security framework for IoTs using identity based cryptography. In: 2016 International Conference on Advances in Computing, Communications and Informatics (ICACCI), Sept. 21–24, 2016, Jaipur, India, pp. 880–886 (2016)
12. Kim, M.S., Kim, S.R., Kim, J., Yoo, Y.: Design and implementation of MAC protocol for Smart-Grid HAN environment. In: Proceedings IEEE 11th International Conference on Computer and Information Technology CIT 2011, pp. 212–217 (2011). https://doi.org/10.1109/CIT.2011.78
13. Pawar, P.M., Nielsen, R.H., Prasad, N.R.: Behavioural modelling of WSN MAC layer security attacks: a sequential UML approach. J. Cyber Secur. Mobil. 65–82 (2012)
14. Babar, S., Stango, A., Prasad, N., Sen, J., Prasad, R.: Proposed embedded security framework for Internet of Things (IoT). In: 2015 2nd International Conference on Wireless Communication, Vehicular Technology, Information Theory and Aerospace & Electronic Systems Technology (Wireless VITAE 2011), pp. 1–5 (2015). https://doi.org/10.1109/WIRELESSVITAE.2011.5940923
15. Tarkowski, M., Rzymowski, M., Kulas, L., Nyka, K.: Improved jamming resistance using electronically steerable parasitic antenna radiator. In: 17th IEEE 17th International Conference on Smart Technologies EUROCON 2017 Conference Proceedings., no. July, pp. 6–8 (2017). https://doi.org/10.1109/EUROCON.2017.8011161
16. Hamza, T., Kaddoum, G., Meddeb, A., Matar, G.: A survey on intelligent MAC layer jamming attacks and countermeasures in the context of a survey on intelligent MAC layer jamming attacks and countermeasures in the context of WSNs. no. July (2016)
17. Qiu, Y., Ma, M.: A mutual authentication and key establishment scheme for M2M communication in 6LoWPAN. IEEE Trans. Ind. Inform. **12**(6), 2074–2085 (2016)
18. Namvar, N., Saad, W., Bahadori, N., Kelley, B.: Jamming in the internet of things: a game-theoretic perspective. In: 2016 IEEE Global Communications Conference GLOBECOM 2016-Proceedings (2016). https://doi.org/10.1109/GLOCOM.2016.7841922
19. Abdalzaher, M.S., Seddik, K., Elsabrouty, M., Muta, O., Furukawa, H., Abdel-Rahman, A.: Game theory meets wireless sensor networks security requirements and threats mitigation: a survey. Sens. (Switzerland) **16**(7), 22–27 (2016). https://doi.org/10.3390/s16071003

Chapter 4
Defense Mechanism for Reactive Jamming Attack in IoT Networks

This chapter provides both individual and combined solutions for reactive jamming attacks. It proposes an efficient defense mechanism for IoT called Countermeasure Detection and Consistency Algorithm (CDCA) for reactive jamming attacks in IoT networks. CDCA detects an attack in the network and cures it. It checks the channel signal strength (to determine the variance in channel signal strength during and after transmission in the presence of jamming) and location consistencies of nodes (to determine location information of legitimate neighboring nodes with their transmission/threshold value) based on a threshold value. Reactive jamming simulations were carried out to compare the CDCA performance to contemporary schemes based on performance metrics.

4.1 The Proposed CDCA for IoT

There are many existing security countermeasures for jamming attacks in the domain of IoT using different defense techniques. Securing IoT networks from jamming attacks is not a new problem; it has been widely studied in the current literature. The goal of a reactive jammer is to prevent an authentic node from sending packets by making the channel always appear busy. In this situation, it is important to monitor the time spent waiting for the channel to be idle, packet signal strength, and node location consistency and compare them to normal traffic time to determine if jamming occurs. To execute this, we introduce CDCA.

The proposed CDCA uses a threshold-based mechanism to detect and cure the attack. In this case, every node maintains some send threshold value and periodically compares the current transmission with the threshold. If it goes beyond that threshold, it understands that an attack has happened and then applies a defensive mechanism. That is, it is a technique that uses a change in the network data/packet to detect an attack in the network and cure it. The major feature of CDCA is that it is a reactive

F. A. Alaba and A. Rocha, *Security Framework and Defense Mechanisms for IoT Reactive Jamming Attacks*, Studies in Systems, Decision and Control 548,
https://doi.org/10.1007/978-3-031-65929-4_4

routing protocol that can sense attack transmission and immediately apply its defense mechanisms against the attacks. It broadcasts parameters like Report Time (i.e., the period during which reports are being sent by each node successively), Attributes (i.e., the physical parameters about which information is being sent), Hard/Higher Threshold (i.e., an absolute value of sensed/detected attribute beyond which node will transmit data to BS) and Soft /Low Threshold (i.e., the smallest sensed value at which the nodes switch on their transmitters and transmit).

The proposed technique also boosts IoT network performance (packet signal strength) and improves connected devices' location consistency checks, saving the network when reactive jamming occurs. The threshold is calculated from Eq. (4.1).

$$\sum_{i=1}^{n} Dx_i = N \tag{4.1}$$

where D is *the* MaxPacket of neighboring nodes, x_i is the distance of each node, n is the total number of nodes, and N is the threshold.

The algorithm uses channel signal strength to check for packet consistency by checking if the data transmission value contradicts the threshold. The node sending a specific threshold to BS is periodically checked and compared with the current value stored at BS after data transmission to determine if there is an attack on the network. If the threshold is greater than the MaxPacket of any particular node, then that node is considered a jammer. But if the threshold is less than the MaxPacket of any particular, the node may or may not be considered a jammer because this can result from lower signal strength during transmission. Also, a jamming attack is detected when the signal strength is high. However, if the signal strength is low, the node may not be jammed. Nevertheless, in this case, there is a need to check and compare the current value received after transmission with the sending threshold on the communicating nodes.

When CDCA detects the jammer node, it notifies all other connected nodes about the jammer node to change all paths coming from the jammer node. As a result, the defense mechanism indirectly removes the jammer node from the network. CDCA completely removed the jammer node by making its path redundant, making it lose its energy and die off.

The CDCA network, attacker assumptions, and algorithm are provided in the next subsections.

4.1.1 CDCA Network and Attacker Assumptions

Below is a list of CDCA assumptions.

a. The network comprises n sensor nodes and is randomly distributed.
b. The capability of each node is the same in terms of functionalities. Each node serves as a gateway (since every node has an IP address) and can message each other.

c. All nodes can communicate with each other either via a single-hop or multi-hop communication.
d. Any node can initiate the jamming attack. A reactive jammer begins to interfere with the channel as soon as it senses an activity on the channel.
e. Both the jamming node and normal nodes are equipped with the same capability, but the jammer node can generate random messages (i.e., random jamming signal).

4.1.2 The Proposed CDCA

The main idea of CDCA is to boost IoT network performance and to save the network when reactive jamming occurs by assigning some data/packet sending threshold to each node in the network. The sending threshold helps to determine a node's ability to send maximum data, which enables the proposed algorithm to detect if there is an attack on the network. The proposed algorithm is presented in Table 4.1.

where X_0 signifies the initial position of the node from point a_0 to b_0, X_n signifies the actual position of the node located from point a_n to b_n, n is the total number of nodes in the network, and D denotes neighbor nodes.

The algorithm involves two phases. In the first phase, the sending node chooses the data-sending threshold from the BS based on the value derived from Eq. 4.1 and checks for signal strength consistency of every participating node. The signal strength consistency provides a feasible way of estimating the distance between participating nodes, as they do not need additional hardware but a radio transceiver. All nodes can

Table 4.1 Flow of CDCA

Algorithm 4.1: CDCA Consistency Check

Required

n, MaxPacket(D): D ∈ Neighbor nodes

Input: Threshold value, a_0, b_0, a_n, b_n;

Output: Dist, ΔThreshold value

```
If (MaxPacket(D) < ThresholdPacket) Then
        Signal Strength (SS) is checked for consistency
        Check the sending threshold value
          Check channel = SS consistency (SS, MaxPacket(D))
Elseif Check channel = false Then
          Jamming occurred
  Endif
If (MaxPacket(D) > ThresholdPacket) Then
        X0 = (a0, b0) = node_location
          Xn = (an, bn) = find_location (n)
          Path Analysis = dist (X0, Xn)
        Check channel = Sent packet
Elseif Check channel = false Then
         Jamming occurred
Endif
End.
```

send data after some interval. The algorithm can keep track of the amount of data sent by each node because it runs on all participating nodes.

The next phase is the checking phase (location consistency), based on the data-sending threshold. The location consistency checks for any change in the threshold of legitimate neighboring nodes. It is assumed that every node in the network is in one of three states: normal state (i.e., non-attacker node), suspicious state (i.e., suspected attacker node), and attacker state (i.e., jammer node). Initially, all nodes are in a normal state and send data to each other using single-hop or multiple-hop communication mode. Among all the nodes, a node is selected based on memory space to be the head node (because different IoT have different memory space and allocation). The head node (BS) stores the threshold of all other neighbor nodes. When the receiving node gets more data than the threshold from a certain source node, the receiving node updates the state of that node to a suspicious state. A path analysis is carried out on the node in a suspicious state. The analysis is carried out in two phases depending on whether the nodes use single-hop or multi-hop communication for exchanging data. If the suspicious source node uses a single-hop, then it is easier for the algorithm to detect the attacker by just performing a single-hop path analysis. If the suspicious node uses multiple hops, the path analysis algorithm checks each hop and the packets all nodes along the path transmits. Once the number of packets produced by the node/nodes is more than the normal send data (in case of multi-hop, a cumulative threshold is calculated using equation (4.1)), then the node is considered to be a jamming node and it is converted to jamming state by the algorithm.

The algorithm decides that a node is in a normal state when the sending threshold packet value is the same with the MaxPacket. In order to determine normal state/ data, the following rules are followed:

a. If MaxPacket is lesser than ThresholdPacket, then the node malicious level is assigned as high.
b. If MaxPacket equals ThresholdPacket, then the node malicious level is assigned as normal. At this level, normal data is received.
c. If MaxPacket is greater than ThresholdPacket, then the node malicious level is assigned as low.

The MaxPacket denotes the maximum packet value from the corresponding entries of each node in the network in every 1s. Finally, the algorithm removes the jamming node by informing other nodes about the detected jammer node and changing the network's communication path. The jamming node is removed from the network by making its path redundant, so the jammer node continuously emits the limited energy it has. Figure 4.1 illustrates how the algorithm works.

The network states indicating how reactive jamming attacks the network and how it is removed from the network are illustrated in Fig. 4.2.

Figure 4.2 shows the IoT network in three states, i.e., in a normal state, attacking stating and curing stating using CDCA. The network state diagram is in a two-dimensional matrix, i.e., time transmitting and time on the channel (the channel can also be "idle"). Data transmission always starts from the top-left state, implying that

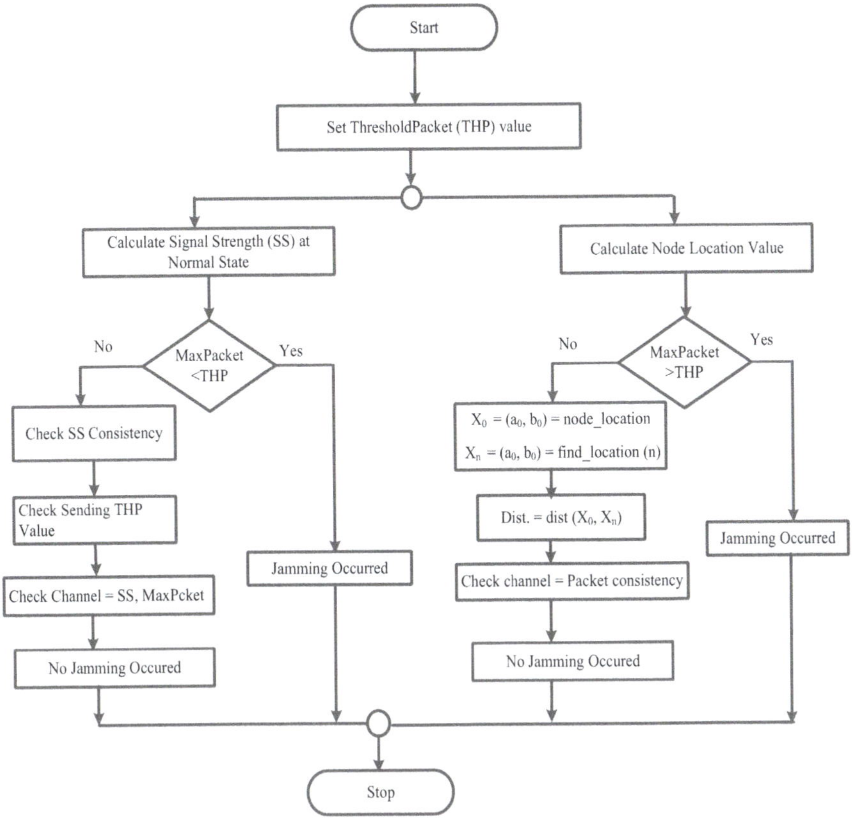

Fig. 4.1 CDCA flow chart

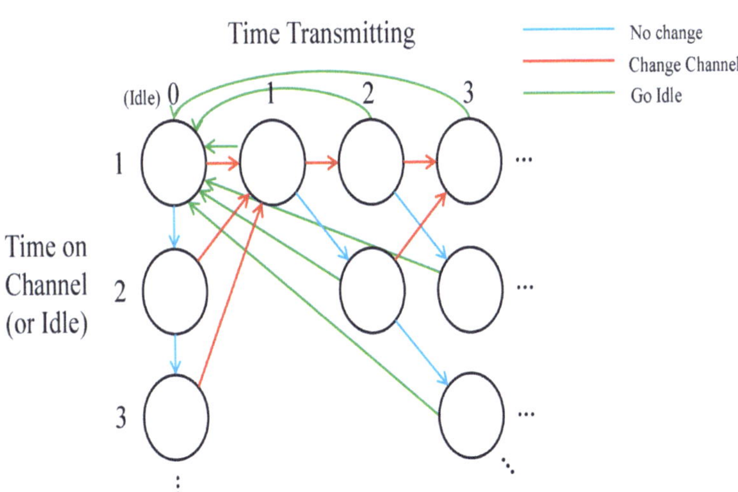

Fig. 4.2 IoT network states with CDCA and reactive Jamming

the network is idle for a one-time step. However, the sending/transmission node can either decide to remain idle or transmit data by changing the channel (i.e., it then starts transmitting from an idle state). If the sending/transmission node chooses to change channel, then in the next step, it must also decide whether to remain on that communication channel or change to a new channel, which we will assume is chosen randomly from a list of available communication channels. It should be noted that the result of each action is deterministic.

A detailed working process of the CDCA in curing reactive jamming attacks is illustrated in Fig. 4.3.

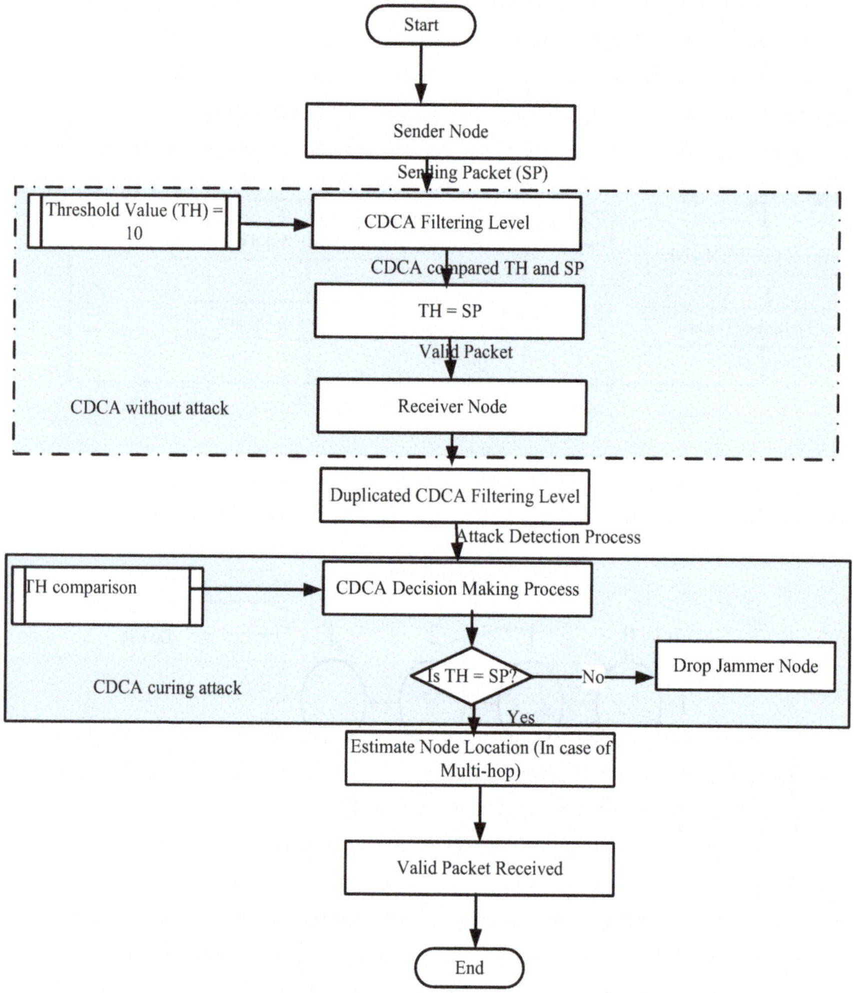

Fig. 4.3 CDCA working process

4.2 Simulation of CDCA

This section provides the implementation of CDCA using the Cooja simulator [1] under varying traffic intervals, number of malicious nodes, realistic conditions, and random mobility.

4.2.1 *Implementation Details*

The Cooja simulator is used to perform the implementation of all the attacks. The parameter for the simulation is based on the IEEE 802.15.4 radio standard. A total number of 100 nodes were placed randomly. The initial energy, idle power, receiving power, sleep power, and transmission power are 100 J, 31mW, 35 mW, 15 μW, and 31mW, respectively, and the threshold value for each node is set to be 10. These numbers provide the basis for properly evaluating jamming attacks [2]. The MAC protocol is Contiki MAC, and the routing protocol [3] used is the ad-hoc routing protocol. The simulation was run 50 times, and the results were compared with the state-of-the-art countermeasures.

The attacks are simulated under the following conditions:

a. **Simulation performed under varying traffic intervals**: This is important to measure and ascertain the performance of jamming attacks in the network and its countermeasures in different traffic situations. The traffic interval ranges between 1 to 10 s, where 1 s is fast traffic, and 10 s is slow traffic.
b. **Simulation performed under various malicious nodes**: This set of simulations is used to analyze the impact of a network attack and its countermeasures by accumulating all malicious objects. Nodes 1, 3, 5, 7, and 15 are malicious nodes. The traffic interval is set to 1 s (the fastest traffic in the network).
c. **Simulation performed under a realistic condition**: A realistic situation is where nodes transmit information at different intervals. The traffic interval was randomly varied between 1 to 10.
d. **Simulation performed under random mobility**: The simulations provide a more realistic CDCA behavior. In this scenario, random mobility is added to each node, and traffic interval is varied between 1 and 10 s. Additionally, mobility speed varies from 1 km/hr to 25 km/hr.

4.3 Results and Discussions

This section compares the performance of the proposed algorithm to the existing TJC countermeasure [4] and the traditional approach [5] for detecting reactive jamming. The results of the average energy consumption, delay, and network throughput are presented in this section.

4.3.1 Performance Results by Varying Interval

The results show an improvement in energy consumption at the nodes, traffic delay, and network throughput when the proposed CDCA is introduced under reactive jamming attack situations compared to the existing algorithm and the traditional approach for detecting reactive jamming. The proposed algorithm analyses the network to detect the jamming attack and improve packet consistency. It also reduces the attack's impact through a separation or isolation process, i.e., it removes the jammer node from the network. The jammer node is technically removed from the network by making it redundant. The redundancy kills the jammer due to excessive emission of the radio signal, which in turn disconnects or removes it from the network. The result for average energy consumption under varying traffic interval is provided in Fig. 4.4.

Figure 4.4 shows the average energy consumption carried out under varying traffic intervals. The result shows that the average energy consumption in the presence of reactive jamming using the traditional approach from interval 1s to 5s is approximately 15%, and the percentage continues to increase with an increased interval. But when the proposed algorithm is introduced in the network simultaneously, the energy consumption is reduced to around 3%, compared to the existing TJC approach that consumes about 11% energy. The proposed algorithm improves energy efficiency because reactive jammers were detected and isolated from the network, thus reducing the energy consumption level due to reactive jamming.

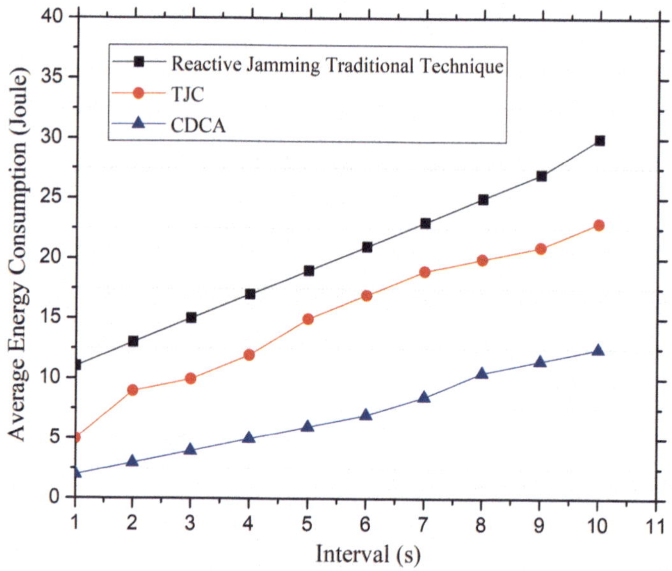

Fig. 4.4 Ave. energy consumption for CDCA, TJC, and reactive jamming traditional technique under varying traffic interval

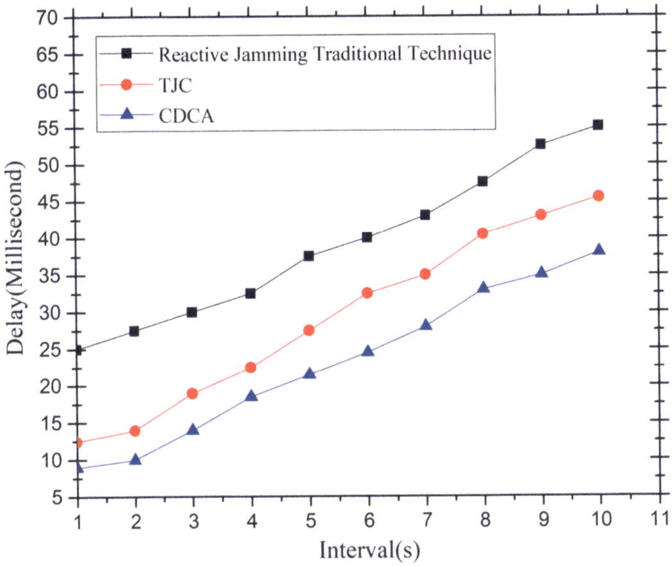

Fig. 4.5 Delay for CDCA, TJC, and reactive jamming traditional technique under varying traffic interval

The result for the network delay under varying traffic intervals is provided in Fig. 4.5.

Figure 4.5 shows the network traffic delay under varying intervals. In a reactive jamming condition, the traffic delay using the traditional approach at interval 1 s is around 25%. Still, when CDCA is applied in the network simultaneously, the traffic delay decreases by almost 9%, 3% less than TJC. The percentage of traffic delay keeps increasing in proportion to the increase in the interval, as shown in Fig. 4.5, because CDCA detected, stopped, and separated/removed the jamming node from the network. In addition, isolating the jamming node indirectly eliminates traffic jams in the channel, which helps to reduce the delay.

The result for the network throughput under varying traffic intervals is provided in Fig. 4.6.

Figure 4.6 shows the network throughput under varying traffic intervals. The result shows that the traditional technique degraded the network throughput to about 42% at an interval of 1 s in a reactive jamming condition. However, the proposed algorithm improves the network throughput to approximately 86, 10% higher than the existing algorithm at the same interval of 1 s. The result shows that the network throughput performance decreases as the interval increases. The reactive jamming condition kept the communication channel busy, thus affecting the network throughput due to a large amount of pending traffic incurred on each node.

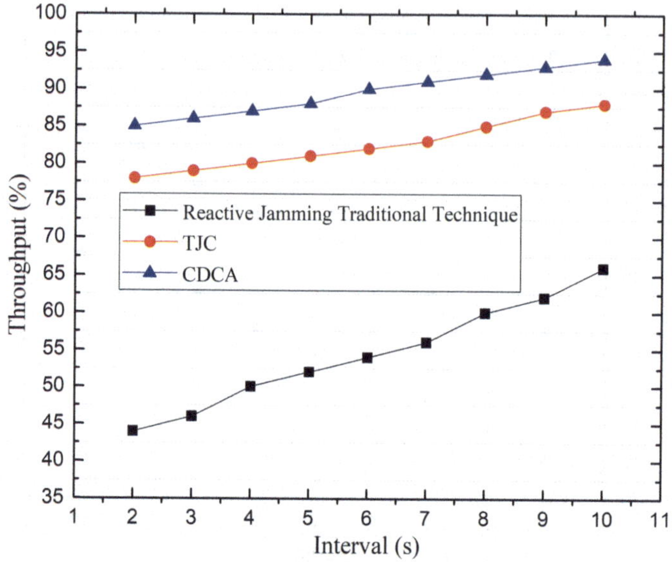

Fig. 4.6 Throughput for CDCA, TJC, and reactive jamming traditional technique under varying traffic interval

4.3.2 Performance Result Under Malicious Nodes

In this situation, the number of nodes increased from 1 to 16, but the malicious nodes were set to 1, 3, 5, 7, and 15. The number of nodes is increased to provide a more realistic/accurate analysis of CDCA when there is an increase in the amount of jamming in the network. Figure 4.7 shows the average energy consumption on the sensor nodes.

The average energy consumption of the sensor nodes under a different number of malicious nodes is shown in Fig. 4.7. The result shows that there is an increase in the level of energy consumption when there are malicious nodes. The energy consumption in a reactive jamming situation using the traditional approach at number 1 and 5 malicious nodes is around 10 and 20%, respectively, increasing to approximately 47% at malicious node 15. The energy consumption level dropped to around 2% and 8% at malicious nodes 1 and 5, respectively, and dropped to 22% at malicious node 15 when the proposed CDCA was applied. This is better than TJC, which consumed about 4 and 14% at malicious nodes 1 and 3, respectively, and increased to around 33% at malicious node 15. CDCA saves energy because its detection mechanism reduces the jamming node energy consumption level by not sending data when an attack is detected. The packet traffic delay performed under varying numbers of malicious nodes is provided in Fig. 4.8.

Figure 4.8 elucidates delay under varying numbers of malicious nodes. The result shows that the traffic delay in a reactive jamming condition using the traditional technique is around 24% at malicious node 1. Delay increases as the number of

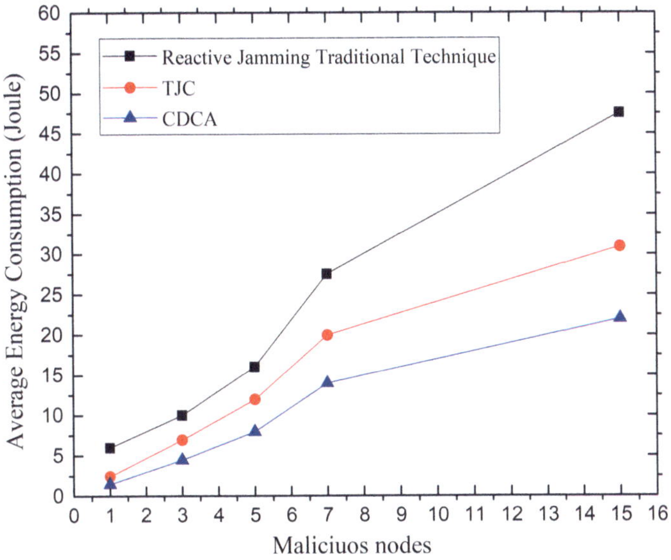

Fig. 4.7 Ave. energy consumption for CDCA, TJC, and reactive jamming traditional technique under malicious nodes

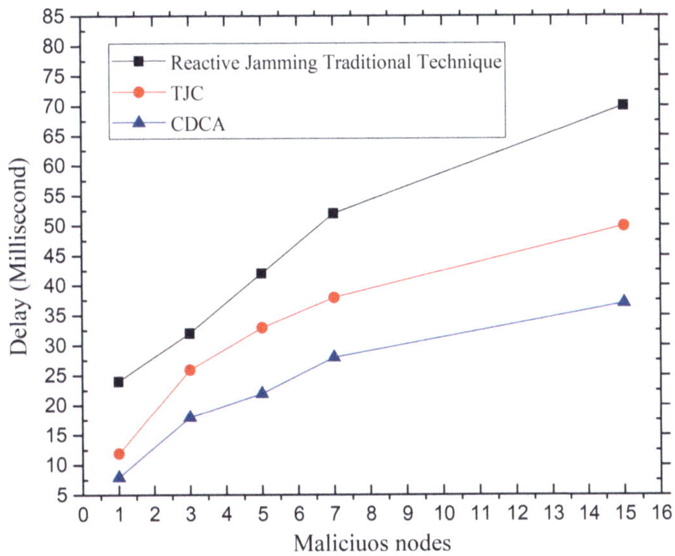

Fig. 4.8 Delay for CDCA, TJC, and reactive jamming traditional technique under malicious nodes

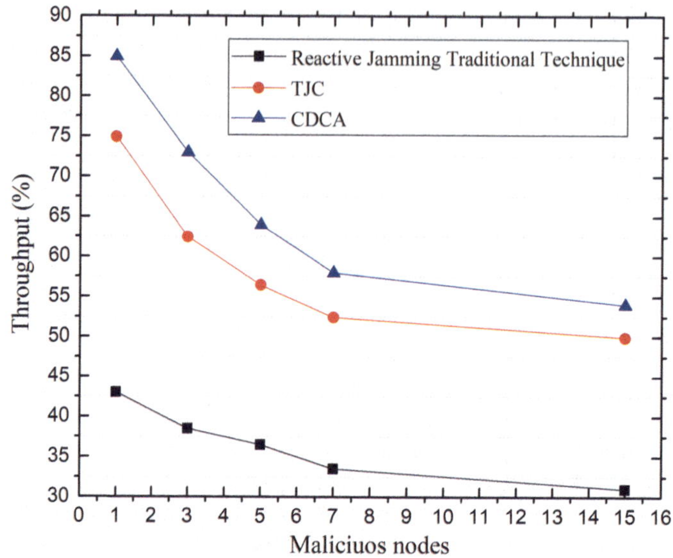

Fig. 4.9 Throughput for CDCA, TJC, and reactive jamming traditional technique under malicious nodes

malicious nodes increases. For example, at malicious node 15, the delay increases to around 70%. However, CDCA helps to reduce the traffic delay to about 8% and 38% at malicious nodes 1 and 15, respectively, compared to TJC, with approximately 12% and 50% traffic delay for the same malicious nodes. The proposed detection algorithm can reduce traffic delay by reducing the channel waiting time.

The results of the network throughput performed under varying number of malicious nodes is presented in Figure 4.9.

Figure 4.9 shows the result of the network throughput as the number of malicious nodes increases. The result shows that the network throughput when the traditional technique is used is about 43% at malicious node 1 and decreases to around 31% at malicious node 15 in a reactive jamming condition. The network throughput is improved to approximately 85% at malicious node 1 and decrease to about 54% at malicious node 15 when CDCA is applied, which is 10 and 5% better than TJC at malicious nodes number 1 and 15, respectively, because CDCA provides rapid channel accessibility to nodes in the presence of reactive jamming.

The performance of CDCA is different for different numbers of malicious nodes due change in attack behavior of each malicious node and the result of each performance metric differs. Also, we look at node 1 and 15 in order to understand the impact of the attack at the entry level (i.e., node 1) and at the peak (i.e., 15) in the network. Also, the results in this condition shows that the proposed algorithm improves network performance. It detected multiple jamming on a single path, which proves that CDCA is efficient and that it can perfectly cure the attacks.

4.3.3 CDCA Performance Result Under Realistic Conditions

The realistic condition simply means that packets were transmitted randomly in the network among nodes between intervals, i.e., each node sent packets at different times. The importance of simulating under the realistic condition is to gain more understanding on CDCA performance in the presence of a reactive attack. The results of the CDCA, TJC and the traditional technique in a realistic condition in terms of average energy consumption, packet traffic delay, and network throughput are shown in Figs. 4.10, 4.11 and 4.12 respectively.

Figure 4.10 shows the result of the average energy consumption in a realistic condition. The result indicates that the energy consumption in reactive jamming using the traditional technique at node number 1 is around 13% and increases to about 55% at node number 15 in a realistic condition. However, the level of energy consumption decreases to about 4% and 29% at malicious nodes 1 and 15, respectively, and the network performance improves when CDCA is introduced to the network in a realistic condition. Moreover, energy consumption is less than TJC by 4% at node number 1 and 11% at node number 15 under the same condition.

The result of the traffic delay in a realistic condition is presented in Figure 4.11.

Figure 4.11 shows the traffic delay in a realistic condition. The delay of the reactive jamming is approximately 35% at node number 1, and it increased to about 80% at node number 15 in a realistic condition when the traditional technique is used. But when the proposed CDCA is applied, it reduced the delay to about 10% at node number 1 and around 44% at node number 15 in a realistic condition, which is better

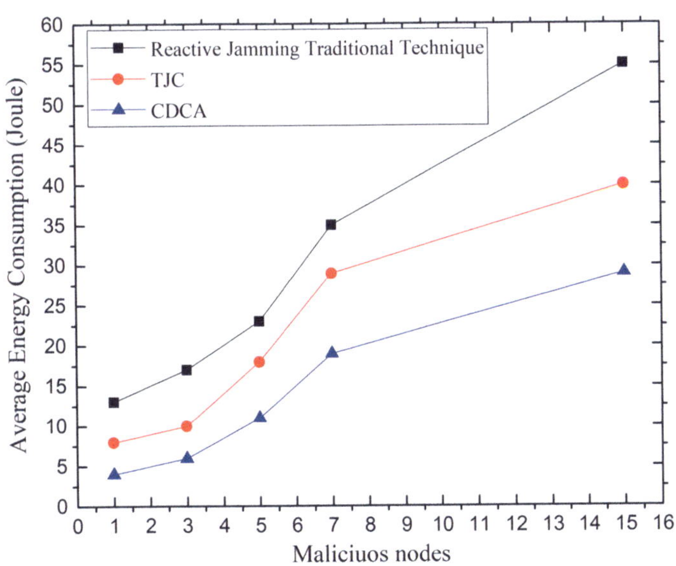

Fig. 4.10 Ave. energy consumption for CDCA, TJC and reactive jamming traditional technique under realistic condition with malicious nodes

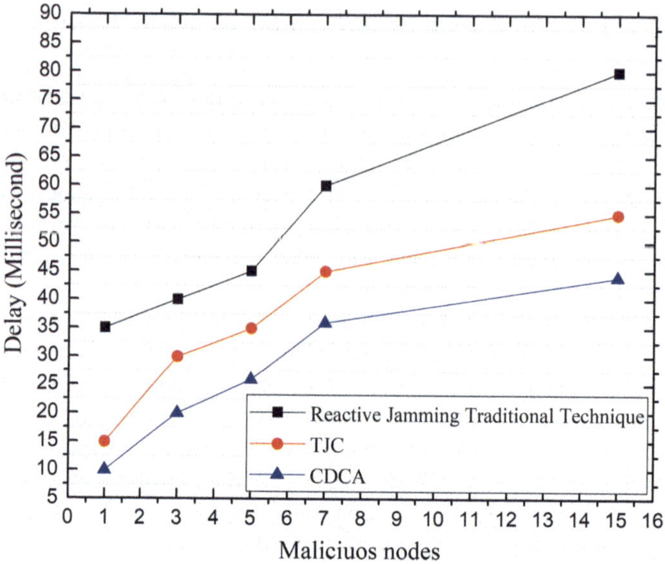

Fig. 4.11 Delay for CDCA, TJC and reactive jamming traditional technique under realistic condition with malicious nodes

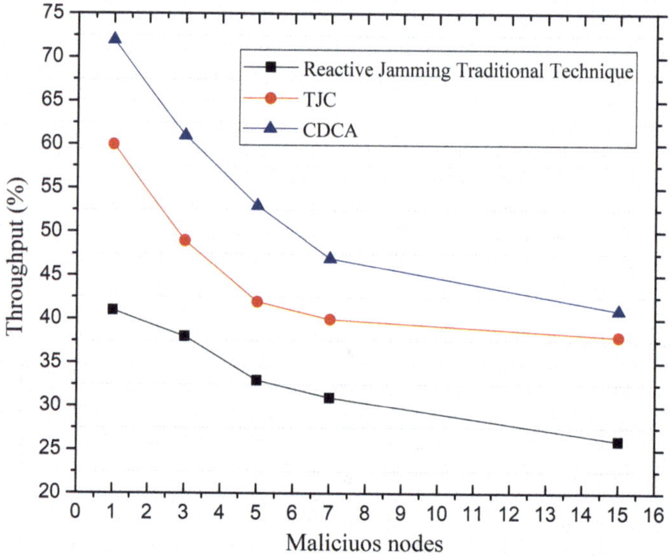

Fig. 4.12 Throughput for CDCA, TJC and reactive jamming traditional technique under realistic condition with malicious nodes

than TJC by about 5% at node number 1 and 11% at node number 15 as shown in Fig. 4.11.

The result of the network throughput in a realistic condition is presented in Figure 4.12.

Figure 4.12 shows the result of the network throughput in a realistic condition. The throughput is around 41% at node number 1 and decreased to about 26% at node number 15 in reactive jamming under realistic condition using the traditional technique. CDCA improves the network throughput to around 72% at node number 1 and decreased to about 41% by node number 15 in a realistic condition. This is 12% better than TJC at node number 1 and also 3% better than TJC at node number 15 in a realistic condition as shown in Fig. 4.12.

4.3.4 Performance of CDCA Under Mobility

The simulation was performed in order to test for the flexibility of CDCA's countermeasure in IoT network. The random waypoint mobility model [6] is considered in the simulation. The measurement also considered both random mobility and random traffic between nodes. The result and measurement of average energy consumption on the sensor nodes, packets traffic delay and network throughput under mobility with different numbers of malicious nodes are presented in Figs. 4.13, 4.14 and 4.15, respectively.

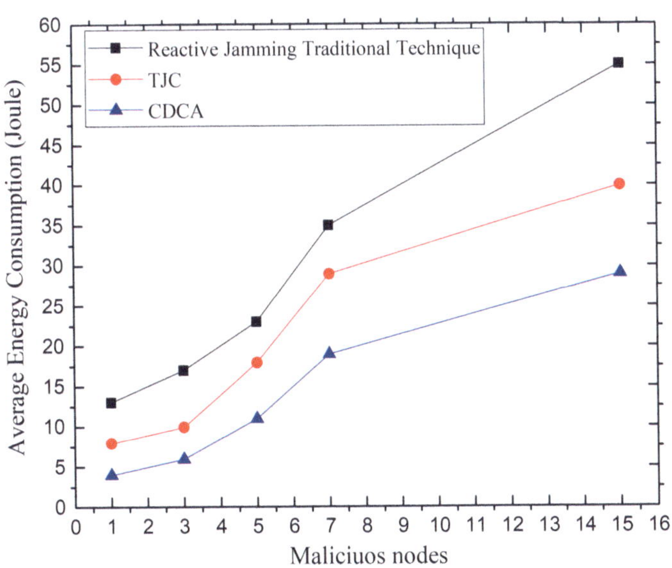

Fig. 4.13 Ave. energy consumption for CDCA, TJC and reactive jamming traditional technique under mobility with malicious nodes

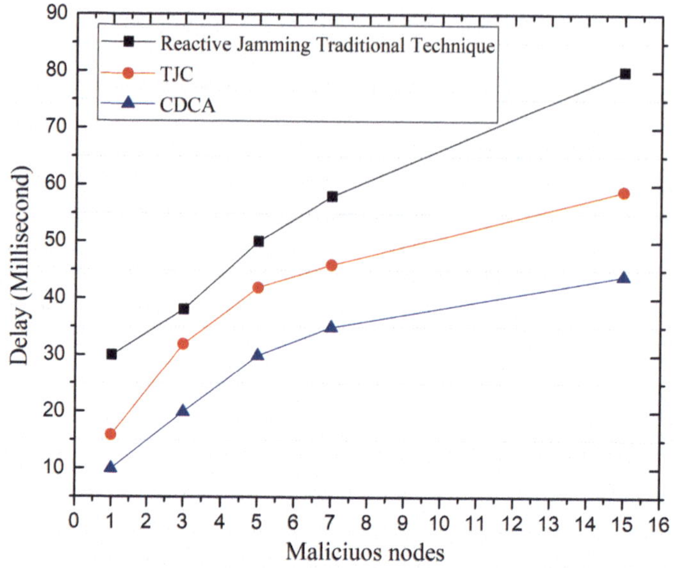

Fig. 4.14 Delay for CDCA, TJC and reactive jamming traditional technique under mobility with malicious nodes

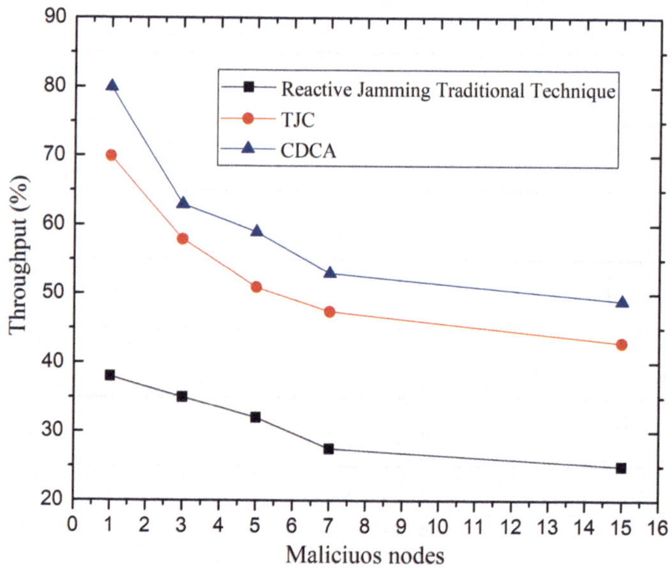

Fig. 4.15 Throughput for CDCA, TJC, and reactive jamming traditional technique under mobility with malicious nodes

The results of the energy consumption under mobility condition are presented in Fig. 4.13. The results show that under mobility and when the traditional technique is applied, reactive jamming average energy consumption at node number 1 is approximately 15% and increases to around 78% at node number 15 as the number malicious nodes increase mobility. The presence of CDCA reduced the energy consumption to about 3 and 23% at node number 1 and 15, respectively, which improves the network performance as the number of malicious nodes increases. It is 6 and 20% less than TJC in terms of energy consumption under mobility.

The result of the traffic delay under mobility condition is presented in Figure 4.14.

Figure 4.14 shows the result of the traffic delay under mobility condition. The result shows that a reactive jamming using traditional approach under mobility causes traffic delay of approximately 30% at node number 1 and increased to around 80% at node number 15, which degrades the network performance. When CDCA is applied in the network, the performance of the network improves by about 10% in terms of delay compared to TJC at node number 1 and with almost 11% of node number 15 under mobility. The introduction of mobility causes an increase in both energy consumption and delay because it takes more time to determine and calculate the threshold value of each node, to detect malicious nodes the network, it needs more energy to scan the pathway.

The result of the network throughput under mobility condition is presented in Figure 4.15.

Figure 4.15 shows the network throughput comparison. The result shows that the throughput for the reactive jamming condition is around 38% at node number 1 under mobility for the traditional approach and decrease gradually to 25% at malicious node 15. However, the network throughput in the presence of the CDCA is approximately 80% under mobility at node number 1 and also decreased 49% at node number 15 as the number malicious increases, which is better than TJC's throughput of around 70% at node number 1 and about 43% at node number 15 under mobility. The reason for the reduced throughput is due to an increase in the time taken to detect the jamming node.

4.4 Conclusion

In this chapter, a novel Countermeasure Detection and Consistency Algorithm (CDCA) for the reactive jamming attack in IoT network was proposed. Using realistic simulation scenarios (e.g., with varying traffic interval, number of different malicious nodes, and random mobility patterns), the performance of the proposed CDCA is evaluated using Cooja simulator. Simulation results demonstrate the performance supremacy of CDCA compared to contemporary schemes in terms of various performance metrics, such as energy consumption, traffic delay, and network throughput. The CDCA reduces the energy consumption and traffic delay, and improve the network throughput compared to contemporary schemes. The main benefit of the novel CDCA is that its defense mechanism supports an increase in the number of jamming nodes in the network. In addition, we provide the result validation using statistical analysis in Chap. 5.

References

1. Thomson, C.: Cooja simulator manual. no. C, pp. 2015–2016, (2016). https://doi.org/10.13140/RG.2.1.4274.8408.
2. Tarkowski, M., Rzymowski, M., Kulas, L., Nyka, K.: Improved jamming resistance using electronically steerable parasitic antenna radiator. In: 17th IEEE International Conference on Smart Technologies, EUROCON 2017—Conference Proceedings, no. July, pp. 6–8, (2017). https://doi.org/10.1109/EUROCON.2017.8011161
3. Li, S., Tryfonas, T., Li, H.: The Internet of Things: a security point of view. Internet Res. **26**(2), 337–359 (2016). https://doi.org/10.1108/IntR-07-2014-0173
4. Babar, S. D.: Security framework and jamming detection for internet of things: a dissertation submitted to the department of electronic system of aalborg university in partial fulfillment of the requirements for the degree of doctor of philosophy. (2015). Available: http://vbn.aau.dk/files/208811916/Thesis_SECURITY_FRAMEWORK_AND_JAMMING_DETECTION_FOR_INTERNET_OF_THINGS.pdf
5. Xu, W., Ma, K., Trappe, W., Zhang, Y.: Jamming sensor networks: Attack and defense strategies. IEEE Netw. **20**(3), 41–47 (2013). https://doi.org/10.1109/MNET.2006.1637931
6. Bettstetter, C., Member, S., Resta, G., Santi, P.: The node distribution of the random waypoint mobility model for wireless ad hoc networks. IEEE Access **2**(3), 257–269 (2013)

Chapter 5
Statistical Results Validation

This chapter presents the results validation of CDCA, TJC and the traditional techniques using statistical models. The statistical model is based on energy consumption, delay, and throughput. The energy consumption, delay and throughput values were derived using regression model with the data collected from the experiments. We used linear regression because of IoT device's constrained nature and heterogeneity to find the significance of the energy consumption, delay, and throughput. Also, its described by the relationship between the variables in order to predict the dependent variable from independent variables.

5.1 Statistical Models

Security is a growing concern as users of small devices in an IoT network are unable to defend themselves against reactive jamming attacks [1–3]. These attacks negatively affect the performance of devices and hinder IoT operations as such, proven statistical models are required to represent the real-life scenario of the data analysis regarding validation. To perform data analysis, the statistical model is used to identify various forms of inferences from the benchmarking experiments. Hence, the purpose of the statistical model is to construct a model that approximates the true structure as accurately as possible through the use of available data [4, 5]. This statistical model is based on three factors: energy consumption, delay and throughput. In realistic settings, many security factors can influence the performance of IoT network. The following elaborates on critical factors closely correlated with the system's performance. The purpose of analyzing these factors is summarized in Table 5.1.

For the statistical analysis model, we used the alteration method to produce numerical values, which contain the average energy consumption, delay, and throughput for each communicating node in the network. These workloads are independent. Moreover, to identify the correlation between the communicating nodes in terms of our

F. A. Alaba and A. Rocha, *Security Framework and Defense Mechanisms for IoT Reactive Jamming Attacks*, Studies in Systems, Decision and Control 548,
https://doi.org/10.1007/978-3-031-65929-4_5

Table 5.1 Summary of the purpose of analysis energy consumption, delay and throughput

Factors	Purpose of analysis
Energy consumption	Compare the average energy consumption of CDCA, TJC, and Traditional techniques under varying traffic intervals and malicious nodes to verify that the proposed algorithm can consume minimum energy compared to the contemporary schemes
Delay	Compare the traffic delay of CDCA, TJC, and Traditional techniques under varying traffic intervals and malicious nodes to show that CDCA can reduce the delay along the channel during transmission compared to the contemporary schemes
Throughput	Compare the throughput of CDCA, TJC, and Traditional techniques under varying traffic intervals and malicious nodes to prove that CDCA can enhance the network throughput in the presence of malicious nodes and under different traffic intervals compared to the contemporary schemes

performance, such as energy consumption, delay, and throughput of the experimental output, we use a regression model to determine the value of the energy consumption, delay, and throughput from the values of one or more variables, so that the dependent variable can be predicted. The dependent variable is energy consumption, delay, and throughput, which mainly depends on the independent variables.

The energy consumption, delay, and throughput predictions can be made from the data collected from the experiments. The energy consumption, delay, and throughput are obtained as numerical values considering the reactive jamming behaviour under different experimental conditions. Linear regression is considered due to independent variables such as size (constrained nature) and the IoT network's number of devices (heterogeneity). Thus, linear regressions are employed to determine the energy consumption, delay, and throughput significance. The regression line depicts the relationship between two variables in a straight line to predict the dependent variable from independent variables. Equation 5.1 shows an example of the mathematical relationship between two variables $x - axis$ and $y - axis$ as the linear regression relationship. The parameters b and m represented the coefficients of the regression model, which are fixed values.

$$y = m * x + b \tag{5.1}$$

For the several independent variable relationships, linear regression is used to predict variables and dependent variables. The main goal of linear regression is to offer a probabilities model that relates a dependent variable to one independent variable. The dependent variable is represented by the time taken for the attacking node to attack, and the independent variables are the amount of data being communicated by the nodes in the network. We used a numerical methods strategy to validate the result of the regression model. Consequently, the generated data from the experiments are used to validate the development of the model. As such, the data split randomly into two: first, to develop the model, and second, to measure predictive accuracy to identify the correlations between the variables.

5.2 Coefficient of Determination

Basically, the measured predicted output corresponds to the actual output accuracy given by the coefficient of determination. It is represented by R^2, which magnitude varies between zero and one. The value with a high number indicates the high accuracy of the prediction. As mentioned earlier, the linear regression models define energy consumption, delay, and throughput. The values of m and b illustrated in Eq. 5.2 will be determined. These parameters cannot be directly measured and have to be estimated. The estimation is performed by collecting data, for example, the total number of nodes and total response time during communication for both normal and malicious nodes. The parameters are estimated by drawing an estimated regression line through the actual data. For any data point, there exists a residual or error of fit represented as r given by:

$$r_i = \alpha_i + \beta_i, \tag{5.2}$$

where α_i the actual node communication time and β_i is the predicted node communication time. The α_i can be less than β_i if the observation shows below the estimated regression line. Using the method of ordinary least squares, the criterion for an accurate prediction is that the sum of the squared residuals be as small as possible, which unexplained variation is represented as:

$$min \sum_{i=1}^{n} r_i^2 \tag{5.3}$$

For any observation I, the execution time α_i deviates from the mean of the original execution times. This deviation is equal to the deviation of the predicted value from the mean plus the residual. The regression explains the predicted value, while the residual remains unexplained. The proportion of the total variation in the energy consumption, delay and throughput is measured by the coefficient of determination R^2 that is explained by the independent variables, which are the number of nodes and the amount data size, as such the R^2 can be defined as:

$$1 - \sum_{i=1}^{n} r_i^2 = \frac{unexplained\ variation}{total\ variation} \tag{5.4}$$

The assessment of the magnitude of R^2 depends on the nature of the communication among nodes that are being analyzed in the network. A R^2 value of 0.05 or higher is said to be a relatively good predication.

5.3 Validation of Results

The performance accuracy of CDCA and the state-of-the-arts approaches were recorded while considering the amount of energy consumption, delay and throughput under different simulation scenario. In order to measure the results of CDCA, validation feature is needed to confirm the quality of the results. This section presents the validation of the proposed algorithm results collected from the experiments for the performance evaluation in Chap. 4. Based on the statistical analysis, the results are related to energy consumption on nodes, traffic delay, and the network throughput performed under different conditions for IoT networks, which show the overall effectiveness of our algorithm. Statistical analysis of our findings via multiple regression models is presented. We use *t*-test results to confirm the significance of the proposed algorithm. The result of the *t*-test shows significant differences between mean values of our performance metrics under varying traffic intervals (i.e., when there is no attack) and in the presence of malicious nodes. The prediction scheme is proving to be accurate, considering the high coefficient of determination values concerning the proposed algorithm.

5.3.1 *Analysis of the Results Regarding Energy Consumption*

Tables 5.2, 5.3, 5.4, and 5.5 illustrates the result of statistical analysis using linear regression and compares CDCA, TJC, and the traditional technique under varying traffic interval, under malicious nodes, under realistic condition with malicious nodes, and under mobility with malicious nodes, respectively.

Tables 5.2, 5.3, 5.4, 5.5 shows the energy consumption comparison of CDCA, TJC and traditional techniques performance measurement based on the result of the experiments conducted under the different simulation conditions. In summary, as

Table 5.2 Comparison between the CDCA, TJC and traditional techniques in terms of energy consumption under varying traffic interval

Interval	Traditional techniques	TJC	CDCA
1	11	5	2
2	13	9	3
3	15	10	4
4	17	12	5
5	19	15	6
6	21	17	7
7	23	19	8.5
8	25	20	10.5
9	27	21	11.5
10	30	23	12.5

Table 5.3 Comparison between the CDCA, TJC and traditional techniques in terms energy consumption under malicious nodes

Interval	Traditional techniques	TJC	CDCA
1	10	4	1.5
3	15	7	4.5
5	20	14	8
7	31	24	14
15	47.5	33	22

Table 5.4 Comparison between the CDCA, TJC and traditional techniques in terms energy consumption under realistic condition with malicious nodes

Interval	Traditional techniques	TJC	CDCA
1	13	8	4
3	17	10	6
5	23	18	11
7	35	29	19
15	55	40	29

Table 5.5 Comparison between the CDCA, TJC and traditional techniques in terms energy consumption under mobility with malicious nodes

Interval	Traditional techniques	TJC	CDCA
1	15	9	3
3	20	17	5
5	32	26	10
7	44	36	17
15	64	43	23

shown from the tables, CDCA consumes low amount energy at every interval both in the presence of malicious nodes and when there is no reactive jamming attack in the network compare to TJC and traditional techniques.

Table 5.6 shows predefined fit function using linear regression for energy consumption in the network. It presents the R^2 value of the attacks being detected during communcation in the network. An R^2, 0.90 and higher are generally considered a good prediction. Considering that even at higher energy consumption level, the value of R^2 is consistently closer to 1 and in most cases above 0.90, this process that the regression model is accurate in its prediction (see Eq. 5.2).

Table 5.6 shows the historical data related to energy consumption using regression model. The table presents mean coefficients, standard errors, and energy consumption time of CDCA and the existing algorithms with 90% confidence interval. The coefficients here represent the mean change in the response variable for one algorithm of change in the predictor variable, whereas holding other predictors in the model constant. The higher values of one variable tend to be related to lower values for the other variable. A positive correlation indicates that higher values of one variable tend to be related to higher values of the other variable. As shown in the table, there is significant in improvement in the network in terms of energy consumption when

Table 5.6 Predefined fit function using multiple regressions regarding energy consumption

Attribute	Value
R^2	0.90
Mean squared error	$4.6e + 3$
Fit parameters	$b = 144.3766 +/- 1.7e + 3$ $m = 19.3 +/- 2.1$
Unformatted fit parameters	$b = 144.37656306548365$ $m = 19.303353662617386$

CDCA is introduced during communication. The performance of CDCA has a direct correlation with good energy reservation on the communicating nodes, meaning that the energy consumption on every communicating node is better using CDCA.

The analysis result of multiple regression of CDCA and the state-of-the-arts approaches for energy consumption in the under of malicious nodes is provided in Table 5.7.

Table 5.7 shows that the significant F is (0.105291), which suggests that the regression procedure estimated by the model is significant at the level of 0.5. The minimum of the coefficients is different from zero, i.e., F (1,3) = 48.2986, $p <$ 0.105291.

The t-value and P-value give an indication of the impact and the significance of each independent variable. The t-value and P-value represent a significant difference between mean of CDCA, TJC and traditional techniques values. The positive t-value of CDCA (0.54249), TJC (−1.00221) and traditional techniques (−0.234214) demonstrate that the energy consumption level on communicating nodes are reduced when CDCA is introduced in the network compare to TJC and traditional techniques. Thus, the energy consumption of CDCA is more significant compared to the state-of-the-arts approaches as presented in Table 5.8.

Table 5.8 shows the coefficients that give intercept and the regression coefficients for each explanatory variable. The coefficient value of the proposed algorithm −0.235266 represents the contestant, which predicted amount enery consumption in comparison with state-of-the-arts approaches. Therefore, CDCA consumes less energy during communication in presence of attacks compare to TJC and traditional techniques.

Lastly, we presents the evaluation comparison of CDCA, TJC and the traditional techniques in Table 5.9.

Table 5.7 Analysis of multiple regression for CDCA and the State-of-the-arts approaches with respect to energy consumption under of malicious nodes

	df	*SS*	*MS*	*F*	Significance *F*
Regression	3	115.9994	38.66648	48.29867	0.105291
Residual	1	0.80057	0.80057	–	–
Total	4	116.8	–	–	–

Table 5.8 *t*-test significance of difference between the CDCA and the State-of-the-arts approaches on energy consumption

	Coefficients	Standard error	*t*-value	*P*-value
Intercept	−0.43339	6.160197	−0.07035	0.955286
Traditional techniques	0.190699	0.814209	−0.234214	0.853535
TJC	−0.93595	0.634545	−1.00221	0.499297
CDCA	−0.235266	2.27703	0.54249	0.683562

Table 5.9 Evaluation comparison: CDCA, TJC and traditional techniques regarding energy consumption

Evaluation	Energy consumption	Coefficients	Standard Error	*P*-value	R^2
CDCA	Low	0.505	1.807	0.214	0.90
TJC	Medium	0.342	1.081	0.788	0.87
Traditional Techniques	High	0.116	0.473	0.761	0.82

Table 5.9 shows the coefficient of determination R^2 "R squared" value of (0.90) from the results gathered from amount of energy consumed during communication. This R^2 is higher than 0.05, which considered a good prediction and proved that the regression model is accurate in its prediction. Therefore, the actual and predicted energy consumption levels by each approach are compared regarding many tasks. The standard error each approach when communicating in the network is also displayed. As shown in Table 5.9, the *estimated coefficients' P-values* are 0.214, 0.788, and 0.761 for the CDCA, TJC, and traditional techniques, respectively, indicating that they are significantly related at the level of 0.05.

5.3.2 Analysis of the Results Regarding Delay

The result of statistical analysis using linear regression that provide the comparison between CDCA, TJC, and the traditional technique under varying traffic interval, under malicious nodes, under realistic condition with malicious nodes and under mobility with malicious nodes are presented in Tables 5.10, 5.11, 5.12 and 5.13 respectively.

Tables 5.10, 5.11, 5.12, 5.13 shows the delay comparison of CDCA, TJC, and traditional techniques performance measurement based on the experiments conducted under different simulation conditions. From the tables, CDCA performs better and reduces congestion along the communication channel at every interval both in the presence of malicious nodes and when there is no reactive jamming attack in the network, compared to TJC and traditional techniques. However, we provide the

Table 5.10 Comparison between the CDCA, TJC and traditional techniques in terms traffic delay under varying traffic interval

Interval	Traditional techniques	TJC	CDCA
1	25	12.5	9
2	27.5	14	10
3	30	19	14
4	32.5	22.5	18.5
5	37.5	27.5	21.5
6	40	32.5	24.5
7	43	35	28
8	47.5	40.5	33
9	52.5	43	35
10	55	45.5	38

Table 5.11 Comparison between the CDCA, TJC and traditional techniques in terms traffic delay under malicious nodes

Interval	Traditional techniques	TJC	CDCA
1	24	12	8
3	32	26	18
5	42	33	22
7	52	38	28
15	70	50	37

Table 5.12 Comparison between the CDCA, TJC and traditional techniques in terms traffic delay under realistic condition with malicious nodes

Interval	Traditional techniques	TJC	CDCA
1	35	15	10
3	40	30	20
5	45	35	26
7	60	45	36
15	80	55	44

Table 5.13 Comparison between the CDCA, TJC and traditional techniques in terms traffic delay under mobility with malicious nodes

Interval	Traditional techniques	TJC	CDCA
1	30	16	10
3	38	32	20
5	50	42	30
7	58	46	35
15	80	59	44

predefined fit function of delay in the network using linear regression in Table 5.14. It presents the R^2 value of the attacks detected during network communication. An R^2 of 0.90 and higher is generally considered a good prediction.

Table 5.14 Predefined fit function using multiple regressions regarding delay

Attribute	Value
R^2	0.93
Mean squared error	6.8e + 3
Fit parameters	$b = 169.3766 \pm 1.7e + 3$ $m = 32.3 \pm 2.1$
Unformatted fit parameters	$b = 169.37656306548365$ $m = 32.303353662617386$

Table 5.14 shows the historical data concerning delay using the regression model. The table presents mean coefficients, standard errors, and delay time of CDCA and the existing algorithms with a 93% confidence interval. The coefficients represent the mean change in the response variable for one algorithm of change in the predictor variable while holding other predictors in the model constant. In addition, a positive correlation indicates that higher values of one variable tend to be related to higher values of the other variable. As shown in the table, there is significant improvement in the network regarding delay when CDCA is introduced during communication.

The analysis result of multiple regression of CDCA and the state-of-the-art approaches for energy consumption in the under of malicious nodes is provided in Table 5.15.

The significant F is (0.230132) as shown in Table 5.15, which suggests that the regression procedure estimated by the model is significant at the level of 0.5. The minimum of the coefficients differs from zero, i.e., $F\,(1, 3) = 9.757459, p < 0.230132$. Moreover, both the t-value and P-value indicate each independent variable's impact and significance. The positive t-value of CDCA (0.12043), TJC (−0.03348), and traditional techniques (−1.283762) prove that CDCA reduces the traffic delay along the communication channel when nodes are communicating in the network compared to TJC and traditional techniques. As a result, the delay of CDCA is more significant compared to the state-of-the-art approaches, as presented in Table 5.16.

Table 5.16 shows the coefficients that give intercept and the regression coefficients for each explanatory variable. CDCA coefficient value is -0.18195, which represents the contestant. This value indicates that the network delay is reduced when CDCA is introduced compared to its contemporary schemes. In conclusion, we present the evaluation comparison of CDCA, TJC, and the traditional techniques in Table 5.17.

Table 5.17 shows the coefficient of determination R^2 "R squared" value of (0.91) from the results accumulated for delay along the channel. This R^2 is higher than

Table 5.15 Analysis of multiple regression for CDCA, TJC and traditional techniques in terms under of malicious nodes

	df	*SS*	*MS*	*F*	Significance *F*
Regression	3	112.9417	37.64723	9.757459	0.230132
Residual	1	3.858303	3.858303	–	–
Total	4	116.8	–	–	–

Table 5.16 *t*-test significance of difference between CDCA, TJC, and traditional techniques on delay

	Coefficients	Standard error	*t*-value	*P*-value
Intercept	−7.48866	4.578122	−1.63575	0.349324
Traditional techniques	−0.426876	0.33252	−1.283762	0.421302
TJC	−0.23087	0.922182	−0.03348	0.978695
CDCA	−0.18195	1.510815	0.12043	0.993698

Table 5.17 Evaluation comparison: CDCA, TJC, and traditional techniques concerning delay

Evaluation	Energy consumption	Coefficients	Standard error	*P*-value	R^2
CDCA	Low	0.505	1.807	0.287	0.91
TJC	Medium	0.342	1.081	0.625	0.84
Traditional Techniques	High	0.116	0.473	0.573	0.80

0.05, which considered a good prediction and proved that the regression model is accurate in its prediction. Therefore, each approach's actual and predicted traffic delay are compared regarding many tasks. The standard error of each approach when communicating in the network is also displayed. As shown in Table 5.17, the *P*-values for the estimated coefficients are 0.287, 0.625, and 0.573 for the CDCA, TJC, and traditional techniques, respectively, indicating that they are significantly related at the level of 0.05.

5.3.3 Analysis of the Results Regarding Delay Throughput

Higher throughput shows that the system can complete more tasks within a given deadline, and the resources can be utilized sufficiently. Many tables are used to demonstrate the findings of our study and compare the results with contemporary methods. Tables 5.18, 5.19, 5.20, 5.21 provides the comparison between the three techniques: CDCA, TJC, and the traditional technique under varying traffic intervals, under malicious nodes, under realistic conditions with malicious nodes, and mobility with malicious nodes are presented in Tables 5.18, 5.19, 5.20 and 5.21 respectively.

The throughput of CDCA, TJC and Traditional Techniques using linear regression over different condition. This predicted throughput is analyzed for all algorithms using same packet in order to find the correlation between them.

Table 5.22 shows predefined fit function using linear regression for throughput. It presents the coefficient of determination; R^2 "R squared" value of the results gathered from tasks processed from all communicating nodes in IoT network. An R^2 0.95 and higher are generally considered a good prediction. Considering that even at higher

Table 5.18 Comparison between the CDCA, TJC, and traditional techniques in terms of traffic delay under varying traffic interval

Interval	Traditional techniques	TJC	CDCA
1	42	76	83
2	44	78	85
3	46	79	86
4	50	80	87
5	52	81	88
6	54	82	90
7	56	83	91
8	60	85	92
9	62	87	93
10	66	88	94

Table 5.19 Comparison between the CDCA, TJC and traditional techniques in terms traffic delay with normal nodes

Interval	Traditional techniques	TJC	CDCA
1	43	75	85
3	38.5	62.5	73
5	36.5	56.5	64
7	33.5	52.5	58
15	31	50	54

Table 5.20 Comparison between the CDCA, TJC and traditional techniques in terms traffic delay under realistic condition with normal nodes

Interval	Traditional techniques	TJC	CDCA
1	35	15	10
3	40	30	20
5	45	35	26
7	60	45	36
15	80	55	44

Table 5.21 Comparison between the CDCA, TJC and traditional techniques in terms traffic delay under mobility with malicious nodes

Interval	Traditional techniques	TJC	CDCA
1	38	70	80
3	35	58	63
5	32	51	59
7	27.5	47.5	53
15	25	43	49

load the value of R^2 is consistently closer to 1 and in most cases above 0.95 this process that the regression model is accurate in its prediction (see Eq. 5.1).

Table 5.22 Predefined fit function using multiple regression regarding throughput

Attribute	Value
R2	0.95
Mean squared error	8.4
Fit parameters	$b = 17.9 \pm 8.8$ $m = 10.1 \pm 1.3$
Unformatted fit parameters	$b = 17.878700000000023$ $m = 10.071249999999997$

Table 5.22 shows the temporal data related to throughput using regression model. The table presents mean coefficients, standard errors, and throughput of 95% confidence interval for CDCA and TJC algorithm. The coefficients here mean the response variable when the predictor variable changed, whereas holding other predictors in the model constant. The higher values of one variable tend to be related to lower values for the other variable. A positive correction indicates that higher values of one variable tend to be related to higher values of the other variable. As shown in the table, there is a significant high throughput when performing tasks using CDCA in IoT network. The performance has a direct correlation with latency, meaning that the system can produce low latency using the CDCA.

Thus, the performance accuracy of the CDCA regarding throughput significantly better than TJC and the traditional techniques. The accuracy of the results is most likely because the datasets used are of approximately equal size. In addition, the data presented in Tables 5.18, 5.19, 5.20, 5.21 were analyzed using ANOVA, and the results are reported in Tables 5.23, and 5.24. Table 5.23 shows that the significant is (0.008). The result suggests that the regression procedure, which is estimated by the model is significant at the level of 0.05. Thus, at least one of the regression coefficients is different from zero. The output of the analysis indicates a statistically significant difference $F(2, 2) = 124.69, p < 0.008 = 124.692$. Therefore, the throughput accuracy of the CDCA is significantly better than the contemporary schemes.

While, as shown in Table 5.24 the *P*-value for the estimated coefficients of x-axis and y-axis, are respectively 0.028 and 0.048, indicating that they are significantly related at the level of 0.05. Therefore, our proposed algorithm framework demonstrates high throughput when performing multiple tasks compare to the contemporary schemes. The *t*-value and *P*-value give an indication of the impact and the significance of each independent variable.

Table 5.23 Analysis of multiple regression for throughput of CDCA and the contemporary schemes

	df	SS	MS	F	Significance F
Regression	2	39.682	19.841	124.692	0.008
Residual	2	0.318	0.159	0.000	0.000
Total	4	40.000	20.000	124.692	0.008

Table 5.24 *t*-test significance of difference between CDCA and the contemporary Schemes with respect to the throughput

	Coefficients	Standard error	*T*-value	*P*-value
Intercept	−6.019	1.007	−5.977	0.027
CDCA	0.852	0.128	5.883	0.028
TJC	−0.540	0.122	−4.420	0.048
Traditional Techniques	−1.19639	−1.303557	−0.91779	0.96158

Table 5.24 reveals that *P*-value is less than 0.05 for proposed algorithm and the contemporary schemes. Since the *P*-value is less than 0.05, the result is accepted at the 5 percent level of significance. Therefore, it is concluded that there no significant difference between proposed algorithm and the contemporary schemes when minimum number of nodes are communicating. Furthermore, Table 5.24 shows the coefficients that give intercept and the regression coefficients for each explanatory variable. The intercept value -6.019 represents the contestant which predicted throughput time. The finding of the study also reveals that the perception of CDCA, TJC and the traditional techniques with respect to throughput is different. A coefficient score of CDCA is 85.2.

Finally, we compared the results of the experiments and the statistical modeling. Table 5.25 provides a comparison of the statistical analysis of throughput results. As illustrated in the table, the results of all the algorithms and statistical analysis show significant high throughput in using CDCA. The average of the high throughput of our proposed algorithm is significant 85.20% when introduced in the network. In addition, 0.852 measures the coefficient. Such, approximately and strong support of findings advocate reliability and validity of the throughput.

Moreover, the table shows the coefficient of determination R^2 "R squared" value of the results. R^2 values are given for the performance of each algorithm under different scenario. A R^2 for CDCA (0.91), TJC (0.86), and traditional techniques (0.76) is captured which is higher than 0.05 that is considered a good prediction and prove that the regression model is accurate in its prediction. Thus, the actual and predicted throughput times are compared to a given number of nodes. The standard error of each algorithm is also displayed. As shown in Table 5.25 the *P*-value of the estimated coefficients of x-axis and y-axis, are 0.145, 0.752 and 0.486, for the CDCA, TJC, and traditional techniques respectively, indicating that they are significantly related at the level of 0.05.

Table 5.25 Evaluation, comparison: CDCA, TJC and traditional techniques in terms of throughput

Evaluation	Throughput	Coefficients	Standard error	*P*-value	R^2
CDCA	High	0.852	0.171	0.145	0.91
TJC	Medium	0.335	0.816	0.752	0.86
Traditional techniques	Low	−0.879	0.841	0.486	0.76

which capture the interaction between nodes and enable nodes to select the best available path for sending data/packets across the network.

After achieving the first two objectives, the findings are used to propose and implement a novel defense mechanism called "Countermeasure Detection and Consistency Algorithm (CDCA)" for the reactive jamming attack in IoT network, and this fulfilled the third objective in Chap. 4.

Finally, the last objective is achieved in Chap. 5, by validating CDCA with the state-of-the-art approaches using energy consumption, delay and throughput as the performance metrics using statistical models. CDCA has been proven to supersede the state-of-the-art approaches based on the listed performance metrics

6.2 Significance

The major contributions of this book are as follows. The results of our extensive literature clearly reveal many security challenges confronting the successful implementation of IoT. The results highlighted some important security issues that needs to be address quickly in IoT domain. Among these issues are; lack of universally accepted IoT architecture, lack of standard authentication scheme that can prevent sensor nodes and RFID from eavesdropping and heterogeneity of IoT devices that leads to jamming on the communication channel as a result of different security protocols used by different communicating devices. To the best of our knowledge, no study/survey of this kind has been carried out prior to this and one of the important findings is that users are willing to adopt and accept IoT if this security challenged mentioned are addressed.

We proceed to propose authentication and capability security framework that is well suitable for IoT devices by integrating AES and GCM authentication techniques which is proven robust authenticating security scheme for IoT devices. The framework provides efficient authentication and encryption standard for IoT devices at low cost, while capability is used as a second line of defense for access control. Moreover, the integration of both techniques makes the framework a challenge-response type that improves computation overhead of IoT. Also, to the best our knowledge, there is no authentication security framework that integrates two different security encryption standards in IoT domain. Next, we adopt UML to develop the activity and behavioural models reactive of jamming attacks. Both models capture all the strategies followed by the malicious attacker in the network, for instance, activity models describe the functional understanding of network, while the behavioural model provide a detailed description of the interactions between nodes sequentially. In addition, game theory was used for formulating reactive jamming attack formation.

Lastly, we propose and implement an efficient defense mechanism that can detect and cure reactive jamming attacks in IoT called "CDCA" and it is evaluated using Cooja simulator. The CDCA was developed based on detail understanding of reactive jamming attacks behaviour in the network and its performance was formally validated and tested with respect to energy consumption on the nodes, traffic delay and network throughput. Our simulation results show the supremacy of CDCA over the existing countermeasures in a realistic scenario under different condition. The CDCA reduces the energy consumption and traffic delay, and improve the network throughput compared to contemporary schemes. Moreover, curing reactive jamming attack is very important in IoT network as earlier stated in previous chapter. Therefore, CDCA has the capacity to cure this disastrous attack that deteriorate IoT network. It is a mechanism the supports an increase in the number of jamming nodes in the network which can increase users' confidence when communicating and also will help to boost success prospects of accepting IoT.

6.3 Future Directions

The focus of this book is to enhance and secure IoT network from malicious attacks, specifically reactive jamming attacks. Efforts were made to address a maximum number of issues, but some are still left unaddressed or discovered during the present work. Below are four research directions that can be used as a starting point for future research:

a. Trust, authorization and privacy: The proposed framework is an authentication-based framework which provide device security and curing of jamming attacks in for IoT network which is an improvement over the existing frameworks. Because among all the existing frameworks discussed in section two (2), there is no framework for device security. However, the proposed framework can also be improved by providing a physical security for embedded IoT devices in trusted platforms. Moreover, the introduction of trust, authorization and privacy at the same time will improve the framework in a better way, as they will provide high-level connections amongst things and services.
b. Session layer: In this work, we focused on securing the physical layer of IoT. Moreover, IoT consist of three different layers; physical, network and application layer. To confirm message integrity across different layers in IoT network, the session provides a secure communication channel between layers. Therefore, it is imperative to extend this work to include session layer protocols that will facilitate the communication among devices across different session in IoT network.
c. In IoT network, identity management is essential to prevent identity theft. This work does not tackle such issue, therefore, there is a need to improve the framework by incorporating dedicated and predefined identity management unity to monitor the connection process of devices.
d. The proposed defense mechanism only cures reactive jamming attacks in IoT network. In the future, this study should be extended to include other types of jamming attacks on IoT networks".

SPRINGER NATURE

GPSR Compliance

The European Union's (EU) General Product Safety Regulation (GPSR) is a set of rules that requires consumer products to be safe and our obligations to ensure this.

If you have any concerns about our products, you can contact us on ProductSafety@springernature.com

In case Publisher is established outside the EU, the EU authorized representative is:

Springer Nature Customer Service Center GmbH
Europaplatz 3
69115 Heidelberg, Germany

The manufacturer's authorised representative in the EU is Springer Nature Customer Service Centre GmbH, Europaplatz 3, 69115 Heidelberg, Germany. If you have any concerns regarding our products, please contact ProductSafety@springernature.com

Printed and bound by CPI Group (UK) Ltd, Croydon, CR0 4YY
23/09/2026
02221546-0015